Soil Chemistry, Fertility and Fertilizers

Soil Chemistry, Fertility and Fertilizers

Rajakumar, G.R.
M.Sc.(Agri), Ph.D
Assistant Professor of Soil Science
Dept. of Natural Resource Management, College of Forestry, Sirsi
University of Agricultural Sciences, Dharwad, Karnataka
e-mail : rajakumar_gr@rediffmail.com

S.V. Patil
M.Sc.(Agri), Ph.D ICP (Holland), PGDAEM,FHAS
Professor of Agronomy
Regional Horticultural Research Extension Centre, UHS Campus
GKVK Post, Bengaluru-560065.
University of Horticultural Sciences, Bagalkot, Karnataka
e-mail: sangappavpatil@gmail.com

SATISH SERIAL PUBLISHING HOUSE
403, Express Tower, Commercial Complex
Azadpur, Delhi-110033 (India)
Phone : 011-27672852, Fax : 91-11-27672046
e-mail : info@satishserial.com, hkjain1975@yahoo.com
Website : www.satishserial.com

Published by :

SATISH SERIAL PUBLISHING HOUSE

403, Express Tower, Commercial Complex, Azadpur, Delhi-110033 (INDIA)
Phone : 011-27672852 Fax : 91-11-27672046
E-mail : info@satishserial.com, hkjain1975@yahoo.com

ISBN 978-93-85055-84-3

Composed, Designed & Printed in India

ಕೃಷಿ ವಿಶ್ವವಿದ್ಯಾಲಯ, ಧಾರವಾಡ
UNIVERSITY OF AGRICULTURAL SCIENCES, DHARWAD

ಡಾ. ಡಿ.ಪಿ. ಬಿರಾದಾರ
ಕುಲಪತಿ

Dr. D.P. Biradar
M.Sc.(Agri), Ph.D.(USA), Post Doc.(USA, Germany & UK)
VICE-CHANCELLOR

Foreword

Soil Science has the wide scope for its application in agriculture and horticulture. The scientific explanation of the subject varies from author to author. The students and the researchers should get the basics of soil knowledge for practical applications in an easy way. Therefore, there is a need for such books. The authors of this book, articulates a vision with the goal of explaining soil, its basics, chemistry and fertility in simple terms. The book entitled **"Soil Chemistry, Fertility and Fertilizers"** will be very much useful for the students and scientists who are engaged in the system.

I wish them "All the best".

Dharwad
2nd May, 2016

(D.P. Biradar)
Vice-Chancellor

YETTINAGUDDA CAMPUS, KRISHINAGAR, DHARWAD-580005, KARNATAKA (INDIA)
Phone : (O) : 091-836-2447783, Fax : 091-836-2448349, Mobile : 94484-95300; Web:www.uasd.edu
E.mail : vc_uasd@rediffmail.com, vcuasd.2010@gmail.com

Preface

Soil chemistry and fertility are the two inter-related disciplines which require basic chemistry and mathematics for understanding the subject easily. Reactions that take place in soil will govern the nutrient uptake by the crop. Management of soil is at most important to get better yields from crops and to save the system with conservation strategies. This book covers brief information on soil physics, soil taxonomy and soil microbiology, besides the main content.

Hopefully, the technical information on soils herein will certainly prove useful to students, scientists and research workers. Efforts are made to provide detailed information on Soil chemistry, fertility and fertilizers besides the quality of irrigation water, soil management and remote sensing. However, any suggestion for improvement of this book will be appreciated and highly welcome. Authors are thankful to the every source of information from where it is drawn, directly or indirectly while preparing this book. Authors also express thanks to the printers 'Satish Serial Publishing House' Delhi for preparation and printing of this book attractively.

May, 2016

G.R. Rajakumar
S.V. Patil

Preface

Soil chemistry and fertility are the two inter-related disciplines which require basic chemistry and mathematics for understanding the subject easily. Reactions that take place in soil will govern the nutrient uptake by the crop. Management of soil is most important to get better yields from crops and to save the system with conservation strategies. This book covers brief information on soil physics, soil taxonomy and soil microbiology, besides the main content.

Hopefully, the technical information on soils herein will certainly prove useful to students, scientists and research workers. Efforts are made to provide detailed information on soil chemistry, fertility and fertilizers besides the quality of irrigation water, soil management and remote sensing. However, any suggestion for improvement of this book will be appreciated and highly welcome. Authors are thankful to the every source of information from where it is drawn, directly or indirectly while preparing this book. Authors also express thanks to the printers Satish Serial Publishing House, Delhi for preparation and printing of this book attractively.

May 2011

G.R. Rajakumar
S.V. Patil

Contents

Chapter 1

Introduction to Soil Science

Definition of Soil

The word soil is derived from Latin word 'solum' which means 'flour'. Soil may be defined as a thin layer of earth's crust which serves as a natural medium for the growth of plants and inhabits. Soil is a dynamic natural body on the upper most surface of the soil containing mineral, organic and microbes with air and water and it supports plants and animals. Soils differ among themselves in some or all the properties, depending on the differences in the genetic and environmental factors. Thus some soils are red, some are black, some are deep and some are shallow, some are coarse-textured and some are fine-textured.

Composition of soil

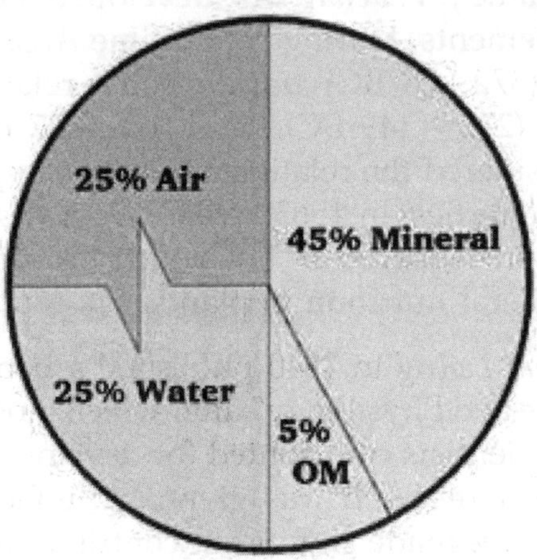

Rocks are the chief sources of the parent materials over which soils are developed. There are: three main kinds of rocks: (i) igneous rocks, (ii) sedimentary rocks and (iii) metamorphic rocks. Liquid magma directly and suddenly cooled and became igneous rocks (IR).. Due to transportation mainly by water these rocks converted to sedimentary rocks (SR) Under high pressure and temperature these rocks converted to metamorphic rocks (MR). Granite and Basalts are the examples for IR, Shale and lime stone are examples for SR and Gneiss and marble are examples for MR.

A study of physical, chemical and biological properties of the soil will give full picture of soil chemistry and fertility in turn the productivity.

Definition of Soil Science

It is the branch of agricultural science which deals with physical, chemical and biological properties of soils in relation to plant growth besides origin, distribution and classification.

History of Soil Science

During 19th century - Shifting cultivation was the main practice because of reduced yields after continuos cultivation of crops. Chinese (300BC) were used to have schematic map of soils of forest area as the basis for taxation. Later, Greeks and Romans employed some sound practice for management of soils.

'Origin of Scientific Study of Soil' can be attributed to German scientist by name *J. V. Leibig* who developed the concept of essentiality of mineral elements. During action time *Aristotle* (384 - 322 BC) and *Jheophrastusl* (372 – 287BC) consider soil in relation to the plant growth. Further *Cato* (232 - 149 BC), *Varo* (116 - 27 BC) and *Virgil* (79 - 19 BC) also considered the relationship between plant growth including forest soil properties. In 1563 *Bernard De Palissy* published a book "A Various Soil in Agriculture" in which he considered soil to be the source of mineral nutrition to plants.

Justus Van Leilrig in 1840 published a book "Chemistry Applied to Agriculture and Physiology" after which modern soil science begins. Later many scientists contributed for development of soil science and importance for research was given. The German scientists *Heyer, Erlenmeyer, Grebe* made significant contribution to the understanding of typical properties soil and importance of litter layer to the stability of forest.*Muller* during 1879 describes the forest humus layer and the influence of biological activity on their development. Later *Ravanna* in 1893 applied soil physical, chemical and biological information to certain forest practices.

Forest soil science was developed very recently in North America. Introduction of new ideas such as sustained yield, reforestation of waste land and establishment of shelterbelts have made people to think about forest soils. The greatest source of the scientific study of forest soil was the publication of text on this subject by *Lutz* and *Chandler* (1946 and 1958), *Wildey* (1958) and recently by *William T. Fisher* (1977 and 1987) on Properties and Management of Forest Soils. *Armson* during 1977 also explained about forest soil, properties and process.

Definition of Forest

It is the place where there is high density of trees with a natural ecosystem. As per FAO forest should have >0.5 ha area, trees of >5 m height and canopy cover of >10% to call an area as the forest.

Approximate major forest area in different parts of the world

Latin America	-	1000 mha
Europe	-	1000 mha
North America	-	500 mha
Africa	-	500 mha
Asia	-	500 mha
Total	-	3500 mha

Different Forest Ecosystem

1. Taiga Forest
    - Pines, spruce
    - Cold / Temperature; more acidic
    - Orders: spodosols, ultisols, oxisols
2. Coniferous Forest
    - America, Canada, England

- Slightly acidic
- Boreal forest is for the largest coniferous forest community
- There are other significant coniferous formation (CF)

i. Transitional CF or subalpine forest in North America and Europe.

ii. Low land CF in US.

These occupy mountains. These soils are leached due to heavy runoff. In higher altitude spodosols and inceptisols are predominant while in lower altitudes. Alfisols and ultisols are prominent.

These forests are seen in America, Canada and England. Here much of land was diverted to agriculture. Great variety of soils is found in these forests. Nutrient recycling is relatively very slow. Has great leaching of behavior hence it is slightly acidic.

3. Mixed Forest
   - South Africa, South America and Newzeland
   - Most suitable for agriculture
   - Soil orders: molisols, insceptisols and ultisols
   - It is the transitional land where the two forest types exist side by side
4. Temperate Broad Leaves Forest
   - Evergreen
   - Japan, South China, parts of Australia
   - Soil orders: inceptisols, ultisols and alfisols
5. Deciduous Forest
   - The forests are evergreen in rainy season and leaf shedding in summer.
   - China, Russia, Australia, USA.
   - The mineral content of deciduous forest is generally higher than coniferous forest.
   - Orders: inceptisols, molisols and alfisols are dominant.

6. Tropical Rain Forest
   - Much of carbon converted to CO_2 because higher decomposition.
   - South East Australia, China, Newzeland.
   - Soil orders: oxisols, ultisols, inceptisols
   - The pH of the soil is slightly low, weathering of parent material is deeper than any other group of soils.

Weathering and intense leaching gradually remove large part of silica from silicate minerals leaving high proportion of Fe and Al oxides. The OM is much higher. Therefore have high biological activity.

7. Temperate Rain Forest
   - Cold, often have rains.
   - Orders: oxisols, ultisols, inceptisols.
8. Tropical Monsoon Forest
   - Less weathered.
   - Ultisols, inceptisols are the soil orders grouped under this.

The type of forest has definite influence on soil formation. Knowledge on this will through a light on the soil type that exists in different parts of the world. If a glance is taken on map of the earth, Ice is the only cover on the surface near Antartic, below which Tundra forests exist (Northern America). Then comes the boreal forest beneath the tundra (northern America, Europe and Russia) followed by temperate grassland (northern central America, parts of Europe and Russia) and temperate forest (parts of China, Russia, Euroipe and East and West parts of Northern America and South coastal parts of Southern. America). Desert and Shrublands exists Australia, China, North western part of India, Pakistan, Afghanisthan, Northern part of Africa and South western part of Northern America. Majority of Indian forest, Africaan forests and South American forest have Savanna. Few Europen parts have Medeteranian Shrubland. Tropical and subtropical forests exist in Australia, Eastern part of China, North eastern part of India, Central and Southern part of Africa, Central and northern part of South America and Southern part of North America.

Branches of Soil Science

The soils can be better understood for management of plant nutrient by study of individual branches, viz., Soil Physics, Soil Taxonomy, Soil Microbiology, Soil Chemistry, Soil Fertility and Fertilizers. In this book, brief basic information on soil physics, soil taxonomy and Soil Microbiology are given. The detailed text is covered on soil chemistry, fertility and fertilizers, besides quality of irrigation water and soil management.

I. Soil Physics: It is the division of soil science that involves the study of physical properties of soil. The chemical and physical relationships among solid, liquid and gaseous materials are affected not only by their own respective properties but also by temperature, pressure and light.

Soil Physical Properties

- **Soil Texture:** the relative proportions of sand, silt and clay (particle sizes) in soil. Clay minerals type determine the textural properties. The layer silicate 1:1 are more in red soils and 2:1 are more in black soils.
- **Soil Structure:** Arrangements of soil particles into aggregates of certain definite pattern.
- **Particle Density (PD):** The weight per unit volume of soil solids, expressed in g/cc, generally 2.65 g/cc, measured by pychnometer.
- **Bulk Density (BD):** The weight per unit volume of soil (volume of soil solids + volume pf pores), expressed in g/cc, generally 1.33 g/cc, measured by Core sampler. In clay soil, the voids (pore space) are more, so have less weight, compared to sandy soil.
- **Porosity:** Pore space is also called as voids. The pore space (PS) consists of the soil volume not occupied by solids.

$$\%PS = (1 - \frac{BD}{PD}) \times 100$$

- **Soil Colour:** A change in colour from adjacent soil indicates differences in soil parent material or differences in pedogenic process. It is measured by Hue (dominant color), chroma (brightness of colour) and value (lightness of colour) Ex: 10 yr 3/4.

- **Soil Water and Air:** exchange of CO_2 and O_2 gases between soil pore space and atmosphere take place continuously.
- **Soil Temperature:** It is an indicator of soil heat absorption capacity. If soil is light coloured it reflects most of the light and absorbs very less. Hence its ALBIDO is less than one.
- **Soil Water Potential :** The moisture held by the soil particles is the sum of different fores, viz., Adhesion and Cohesion, (matric potential), gravitational potential and solute potential (due to salts).
- **Soil Moisture constants:** At a particular tension the water is held by the soil particles known as soil moisture constants, viz., Water Holding Capacity (WHC), Field Capacity, Permanent wilting point, Hygroscopic coefficient and others.
- **Maximum WHC :** It is the moisture content held in the soil both in micro and macro pores where the gravitational water still exists.
- **Field Capacity (FC):** The amount of water retained in the soil after the cease of downward movement of gravitational water from the saturated soil. The potential (tension) is less than 0.33 Bar.
- **Permanent Wilting Pont (PWP):** The amount of water held at potential of 15 Bar.
- **Available Water :** It is the moisture available to plants which can obtained **by** moisture contents at FC minus PWP. The capillary water is held between 1/3 and 31 atmosphere pressure.
- **Hygroscopic Water** It is the vapour type of moisture called as coefficient where the tension varies from 31 atm to 10000 atm.
- **Soil pF** : The free energy is measured in terms of the height of a column of water required to produce necessary suction or pressure difference at a particular soil moisture level. It is the negative logarithm of cm of water column held against the air pressure in soil to give the necessary suction. A porous ceramic cup can be installed in soil with a connection to water pipe fixed with pressure gauge to measure the tension. In dry soils the pF will be 7.0, in wet soils at FC the pF will be 2.54. The capillary water will have pF of 2.5 to 4.5
- **Hysteresis behaviour of Soil moisture :** The moisture content at same suction / tension differs under two wetting and drying

fronts. The moisture content on wetting will be different and the same on drying will be different besides the same suction, which is referred as the hysteresis.

II. Soil Taxonomy: It is derived from two word viz. **Taxa:** Characteristic feature of soil of an area used for classification and delineation (into map) and **Nomy :** means the study. This involves pedology also. It is the division of soil science that deals with the genetic origin, survey and classification of soils as a body in nature. Dockuhaiev (1900), the father of soil science, he was the first to enunciate the fundamental principles of pedology. It is the Division of soil science which deals with the genetic origin, survey and classification of soils as a natural body.

Soil Forming Factors: Soil is a function of climate, topography, organism and parent material and time.

Climate and organism are tae active factors because they supply the energy for formation of soil. Parent material, relief and time are the passive factors because they are the source of soil development.

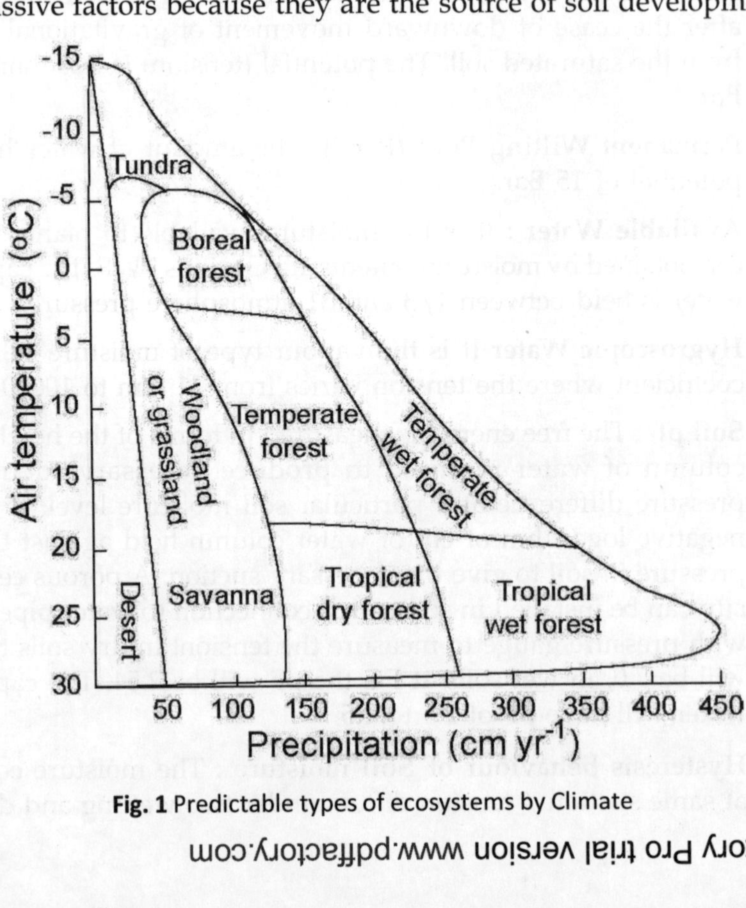

Fig. 1 Predictable types of ecosystems by Climate

a) **Climate:** It includes rain fall, temperature, humidity, wind. Climate is the more important factor that influences soil formation. In arid and semi arid areas it may favour lateritic acidic soil due to intense weathering and leaching of basic cations. Similarly in humid and cool conditions favours formations of podzols or forest soils.

b) **Organism:** The activity of living plants and animals and decomposition of organic waste by microbes have influence on formation of soil and profile development.

c) **Parent material:** It is the most important factor and it is the initial soil materials that determine the soil profile development. It influences soil formation depending upon intensity of weathering and materials they contain.

d) **Topography:** The earth surface slope is called topography. It influences soil formation primarily through soil erosion: soils on the upper slope (top hill) are different from base of the hill.

e) **Time:** The length of the time required for soil formation or horizon development depends upon many interrelated factors like climate, nature of parent material and relief. Under ideal condition the time required for development of soil may be between 100 – 200 years. If the conditions are not ideal the time may be extended up to 1000 years.

Soil was formed during "Pleistocene age". Soil profile undergoes certain pedogenic process. Well formed soil is a result of several processes is collectively called pedogenic process. Fundamental processes according to Simonson (1959) are addition, loss, transformation and translocation.

Fundamental Soil Forming Processes

1. *Humification:* transformation of raw organic matter in to humus
2. *Eluviation (Lessivage):* Analogous to emigration

   Eluviation means washing out. It is the process of removal of constituents in suspension or solution by the percolating water from the upper to lower layers. French scientists use an analogues term lessivage to describe the mechanical movement of clay and Fe oxides from the A horizon without undergoing chemical alteration (A_2).

3. *Illuviation:* Analogues to immigration

   The process of deposition of soil materials (removed from the eluvial horizon) in the lower layer (or horizon of grains) that have the property of stabilizing translocated clay materials is termed as illuviation. The horizon formed by this process leads to textural contrast between E and Bt horizons and higher fine clay to total clay ratio in the Bt horizon.

Specified Pedogenic Process

1. **Calcification:** It is the process of precipitation and accumulation of $CaCO_3$ in some part of the profile. The accumulation of $CaCO_3$ may result in the development of a calcic horizon. (Grasses, Desert shrubs)

$$CO_2 + H_2O \rightarrow H_2CO_3$$

$$H_2CO_3 + Ca \rightarrow Ca(HCO_3)_2 \text{ (soluble)}$$

$$Ca(HCO_3)_2 \xrightarrow[CO_2]{temp} CaCO_3 + H_2O + CO_2$$

(precipitates)

2. **Decalcification:** It is the reserves of calcification that is the process of removal of $CaCO_3$ or calcium ions from the soil by leaching.

$$CaCO_3 + CO_2 + H_2O \xrightarrow[CO_2]{temp} Ca(HCO_3)_2$$

(Insoluble) (soluble)

3. **Pedzoliazation:** It is a process of soil formation resulting in the formation of podzols (spodosols) and podzolic soils. It is the negative of calcification. Apart from calcium, other basis is also removed and the whole soil becomes distinctly acidic.

   Under pines, waxy, material release and react with clay, where cold and humid climate exist.

4. **Laterization:** The term laterite is divided from the word later meaning brick or tile and was originally applied to a group of high clay; Indian soils first observed in the Malabar Hills of Kerala where a monument has been erected in its commemoration. Such soils are also observed in the other states Tamil Nadu, Karnataka, Odisha and Maharashtra.

The laterite refers specially to a particular cemented horizon in certain soils which when dried become very hard and used as bricks for construction of buildings under warm and humid climate. Laterization is the process that removes silica instead of sesquioxides to concentrate in the solum.

5. **Gleization:** The term glei is of Russian origin which means blue, grey or green clay. The gleization is a process of soil formation resulting in the development of a glei (or gley horizon) in the lover part of the soil profile above the parent material due to poor drainage condition, (lack of O_2 and water logged condition prevail). Such soils are called hydromorphic soils.
6. **Salinaization:** It is the process of accumulation of salts, such as sulphates, chlorides of Ca, Mg, Na and K in soils in the form of a salty (salic) horizon. As a result, the accumulation of salts, solonchaks or saline develops.

   (Gangavathi ARS in Karnataka have soils with gradient, non-saline to highly saline).
7. **Desalinization:** It is the removal by leaching of excess soluble salts from the soil profile (that contained enough soluble salts to impair the plant growth) by standing water and improving the drainage condition by installing artificial drainage network.
8. **Solonization:** The (alkalization) process involves the accumulation of Na ions on the exchange complex of the clay resulting in the formation of sodic soils (solonization). All cations in solution are engaged in a reversible reaction with the exchange sites on the clay and organic matter particles.

   X) CaMg 2Na $\rightarrow Ca^{++} + Mg^{++} + 2Na^{+} + x^{-6} + 3CO_3^{2-} \rightarrow Na_2CO_3 + MgCO_3 + CaCO_3$

   (where 'x' represents exchange sites on clay or OM)
9. **Solodization (De-alkalization)**: The process refers to the removal of Na^+ from the exchange sites. The process involves dispersion of clay

   • Ultisols, inceptisols are the soil orders grouped under this.

   $2Na\ x + CaSO_4 \rightarrow Na_2SO_4 + Ca\ x$ (x is exchange sites)
10. **Pedoturbation:** It is the process of mixing of soil. Mixing of clay certain extent takes place in all soils. Common types are:

i. *Faunal pedoturlation:* It is mixing of soil by animals such as ants, earthworms, moles, rodents and man himself.

ii. *Floral pedoturlation:* that forms pits and mounds

iii. *Agrillo pedoturbation:* It is the mixing of materials in the solum by the churning process caused by swell-shrink clays, as is observed in deep black cotton soils of central India.

Different Forest Ecosystem: Rain forest, dry forest, savannas and desert are the different forest ecosystems where the soil profile development differs from each other. In rain forest there is little soil profile development due to climatic factors but more influenced by plant roots and percolating organic matter.

A developed soil will have a well defined profile which is a vertical section of the soil through all its horizons and it extends up to the parent material. The horizons (layers) in the soil profile which may vary in thickness may be distinguished from morphological characteristics which include colour, texture, structure etc. Generally, the profile consists of three mineral horizons - A, B and C. An organic layer 'O' may be found in forest soils with an organic carbon content of more than 10%. The A horizon may consist of sub-horizons richer in organic matter intricately mixed with mineral matter. Horizon B is below A and shows dominance of clay, iron, aluminium and humus alone or in combination. The C horizon excludes the bedrock from which A and B horizon are presumed to have been formed.

Simplified Soil Profile: with Master horizons and Sub horizons

O	Organic	Organic Matter can be seen only in forest soils
A	Mineral	Mineral layer which is fine in texture compared to B horizon top layer
B	Bineral	Lies in between A & C horizon and has both the feature of subsurface A & C
C	Unconsoli	Unconsolidated rock fragments dated
R	Consoli	Consolidated (huge mass of rock)/ parent material dated huge mass of rock

In most of the areas, including in agricultural soils, the 'o' horizon will be absent. Only in forest soils, the organic horizon can be seen.

Horizons expressed with suffixes :

a - highly decomposed organic matter;

b - buried soil layers;

c - concretions;

e - intermediate decomposed OM;

f - fragipan;

h - illuvial OM;

n - sodium rich;

o - oxides of Fe & Al;

x - high bulk density;

z - salt rich;

t - clay rich

Horizon Characters

a. Surface Horizon: (Epipedons)

H – Histic (more than > 25% Organic Matter, soft)

O – Ochric (Oxidised pale colour)

P – Plaggen (disturbed by man)

U – Umbric (more P_2O_5 and Percentage base saturation)

M – Mollic (grassland, more Organic Matter)

A – Argillic (cultivated soil with more clay)

Kandic – clay accumulated;

Sombric – OM accumulation;

Duripan-hard pan;

Fragipan – Brittle pan;

Spodic – Fe & Al rich

b. Subsurface Horizon: (Endopedons)

Nitric – Na endopedons

Salic – salts endopedons

Argillic – clay rich;

Oxic – highly weathered

Albic – light coloured;

Horizons vertical variability

Soil profile is a result of long-term physical, chemical and biological processes. How a soil develops at a specific site depends on the relative importance of various soil-forming factors (i.e. climate, parent material, topography, time vegetation and fauna including man). Soil formation often involves transport of specific compounds resulting in depletion in some parts of the profile and enrichment in others. For example, due to the chelating power of percolating fulvic acids, Podzols have a strongly acidified, bleached Al and Fe-depleted E horizon at the surface. At depth Fe and Al chelates precipitate, thus forming a spodic B horizon, enriched in Fe, Al and organic C. At the surface, Podzols often develop a layer of acidic, slowly decomposing raw humus (Organic horizon). Other examples are:

1. Translocation of clay minerals, giving rise to a clay-depleted horizon.
2. Accumulation of salts at or near the surface (calcic, gypsic, natric-B horizon), as in Solonchaks and Solonetzes.
3. Intensive weathering and leaching of Si in the humid tropics leads to a ferralic horizon, with more of Fe and Al-oxides and kaolinite (Ferralsols).
4. Peats (Histosols) may form as a result of accumulation of organic matter, due to incomplete decomposition of plant debris, rather than leaching, specially under forests.
5. Gleization: The term *glei* is of Russian origin means blue, grey or green clay. The Gleization is a process of soil formation resulting in the development of a glei (or gley horizon) in the lower part of the soil profile above the parent material due to poor drainage condition (lack of oxygen) and where waterlogged conditions prevail. Such soils are called hydro orphic soils.

Different soil types (catena) may develop on uniform parent material, due to variation in relief, drainage and vegetation. For example, soils with eluvial (leached) horizons generally develop on hillslopes and summits (e.g. Podzols, Luvisols, Acrisols, Ferralsols), while soils in depressions (e.g. adjacent to the stream) may exhibit gleyic properties or accumulation of organic matter (Histosols).

Soil Orders

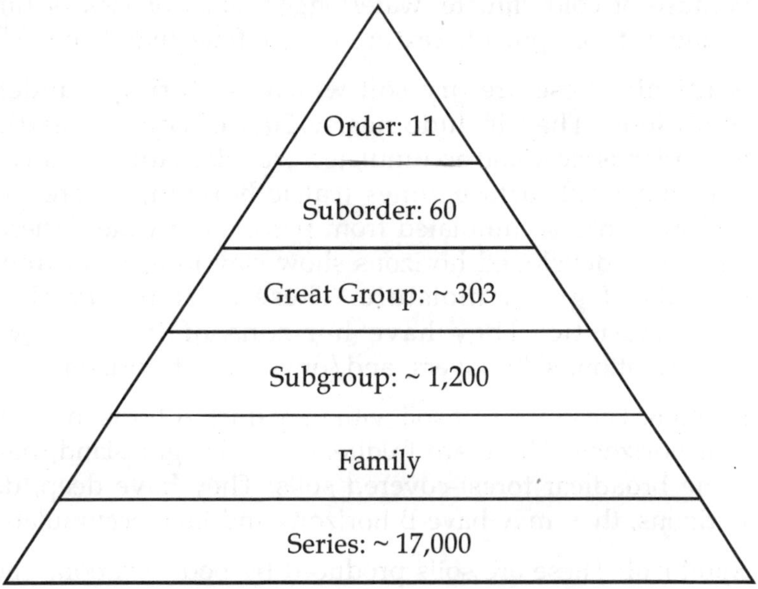

Fig. 2 Triangle of soil orders to series

Study of soil profile is most important to categorize a soil of an area into one group, called soil order. There are 12 soil orders, 60 Suborders, 303 great groups, 1200 subgroups many families and series in Soil Taxonomy.

Soil orders

1. **Entisol:** These are recently formed soils that lack well developed horizons, commonly found on unconsolidated sediments, some have simply an A horizon on top of bed rock. Entisols have a shallow marginal A horizon. River floodplains, volcanic ash deposits and sands are Entisols.
2. **Vertisol:** These are self inverted soils. They tend to swell when wet and shrink upon drying, often forming deep cracks that make surface layers fall in to it. They have shiny slicken sides in the subsurface horizons due to clay rubbing on clay. These soils exist in temperate to tropical climate with distinct wet and dry seasons.
3. **Inceptisol:** These are young soils. They have subsurface horizon formation but show little deviation among horizon. Inceptisols,

in humid regions, have weak to moderated horizon development because of cold climate, waterlogged soils, or lack of time for stronger development. Texture will be finer than loamy sand.

4. **Aridisol:** These are dry soil which are forming under arid conditions. They include nearly 20% of soils on earth. Soil formation is slow and accumulated organic matter is scarce. They may have subsurface zones (calcic horizon) where calcium carbonate has accumulated from percolating water. These soils have well developed horizons show clay movement from past periods of grater moisture. These exist in dry climates. Characteristic: They have horizons of lime or gypsum accumulation, salty layers, and/or A and Bt horizons.
5. **Molisol:** These are soft soil with very thick A horizons and other thin horizons. These are frequently under grassland, but with some broadleaf forest-covered soils. They have deep, dark A horizons, they may have B horizons and lime accumulation.
6. **Spodosol:** These are soils produced by podsolization. They are typical soils of coniferous and deciduous forests in cooler climates. These are typically the sandy, leached soils of cold coniferous forests.They have O horizons, strongly acid profiles, well-leached E horizons, Bh or Bs horizons of accumulated organic material plus iron and aluminum oxides.
7. **Alfisol:** These are soils rich with aluminum and iron. They have horizon of clay accumulation and from where there is enough moisture and warmth from at least three months. Alfisols develop in humid and sub-humid climates, have average annual precipitation of 500-1300 mm. They are frequently under forest vegetation. They have clay accumulation in a Bt horizon, thick E horizon, available water much of the growing season, slightly to moderately acid.
8. **Ultisol:** These are soils that are heavily leached. These are extensively weathered soils of tropical and subtropical climates.They have thick A horizon, clay accumulation in Bt, strongly acid.
9. **Oxisol:** These are soils with oxidized horizons. These are excessively weathered, whereas few original minerals are left unweathered. They develop in tropical and subtropical climates. These are over 3 m deep, have low fertility, have dominantly iron and aluminum clays, and are acid.

10. **Histsol:** These are organic soils. Peet and muck soils come under this. These are consisting of variable depths of accumulated plant remains in bogs, marshes, and swamps.

11. **Andisol:** These are volcanic soil, which tend to be high over 60% in ash (ash, cinder, pumice, basalt)content. These are with bulk densities below 0.9 g/cc. These have dark A horizon, early-stage secondary minerals (allophane, imogolite, ferrihydrite clays), high adsorption and immobilization of phosphorus, very high cation exchange capacity.

12. **Gelisol:** These are permafrost soils the condition where ice (solid/gel) is present in within 200 cm of the ground surface horizons.

Major Soils of India are

1. **Red Soil**: They are wide spread soils in India, red in colour due to diffusion of iron in the profile. Belongs to Alfisol, Inceptisol and Ultisol soil orders.
2. **Lateritic soil**: These are composed of a mixture of hydrated oxides of aluminium and iron with small amounts of manganese oxide. These are found in Assam, Kerala and parts of Karnataka. Belongs to Alfisol, Ultisol and Oxisol soil orders.
3. **Black soil**: These soils contains more clay, a high proportion of Calcium and Magnesium on exchange sites and have a high degree of fertility. Belongs to Vertisol, Inceptisol and Entisol soil orders.
4. **Alluvial soils**: This is the largest and agriculturally important group of soils. Belongs to Entisol, Inceptisol and Alfisol soil orders.
5. **Desert soils**: occur mostly in dry areas and important content is Quartz (Sand). Belongs to Entisol and Aridisol.
6. **Forest and Hill Soils**: high in organic matter content. Belongs to Alfisol soil orders.

Land Capability Classification

The soils are studied and classified according to their use which is termed as Land Capability Classification (LCC).

In this classification, inherent soil characteristic, external land features and environmental factors are given prominent importance.

For this purpose, soil survey is carried out to record the crop limiting factors such as soil depth, topography, texture-structure, water holding capacity, drainage features, followed by evaluation of soil fertility status based on soil testing. The soils are thus classified in to 8 classes, four of which are considered suitable for agricultural purpose (Class I & IV) and Class V to VIII are non-arable lands and can be used for silviculture and forest and those areas need strong conservation measures to protect and develop soils.

III. Soil Microbiology: It is the division of a soil science which deals with microscopic population of the soil, its role in various nutrient transformations and its importance in plant nutrition and crop production. **Soil microbiology** is the study of organisms in **soil**, their functions, and how they affect **soil** properties. It is believed that between two and four billion years ago, the first ancient bacteria and microorganisms came about in Earth's primitive seas.

Soil microbiology consist study of bacteria, actionomycetes, fungi, mycorhizae and yeast, etc., their activities, functions, and role in maintenance of soil status. They are mainly involved in decomposition of soil organic matter, carbon and nitrogen fixation. **Soil microbial activity** that reflects **microbiological** processes of soil microorganisms is the potential indicator of soil quality, as plants rely on soil microorganisms to mineralize organic nutrients for growth and development.

$$\text{Organic Matter} \rightarrow \text{inorganic } NH_4 \xrightarrow{\text{Nitrosomonos}} NO_2 - \xrightarrow{\text{Nitrobactor}} NO_3 - \} \text{ Nitrification}$$

Microbes

→ Other constituents → Easily soluble and chelated products

- When it rains, we get soil smell. It is due to death of actinomycetes. If soil is fertile there will be a good smell of soils.
- Fungi – are the important microbes for degradation of Soil Organic Matter Ex: Trichoderma.

After degradation, the CO_2 is released to the atmosphere and nutrients to the soil.

- Mycorrhiza: Mycile like structure, association between fungi and lower plants.

The scope and importance of soil microbiology, can be understood in better way by studying Soil as a living system, Soil microbes and plant growth, Soil microorganisms and soil Structure, Organic matter decomposition, Humus formation, Biogeochemical cycling of elements, Soil microorganisms as bio-control agents, Soil microbes and seed germination, Biological N2 fixation and Degradation of pesticides in soil.

Soil inhabit diverse group of living organisms, both micro flora (fungi, bacteria, algae and actinomycetes) and micro-fauna (protozoa, nematodes, earthworms, moles, ants). The density of living organisms in soil is very high i.e. as much as billions / gm of soil. Microorganisms being minute and microscopic, they are universally present in soil, water and air. Soil structure is dependent on stable aggregates of soil particles-Soil organisms play important role in soil aggregation.

Organic matter / substances include cellulose, lignins and proteins (in cell wall of plants), glycogen (animal tissues), proteins and fats (plants, animals). Cellulose is degraded by bacteria, especially those of genus *Cytophaga* and other genera *(Bacillus, Pseudomonas, Cellulomonas, and Vibrio Achromobacter)* and fungal genera *(Aspergillus, Penicilliun, Trichoderma, Chactomium, Curvularia)*. Lignins and proteins are partially digested by fungi, protozoa and nematodes. Proteins are degraded to individual amino acids mainly by fungi, *actinomycetes* and *Clostridium*. Under unaerobic conditions of waterlogged soils, methane are main carbon containing product which is produced by the bacterial genera (strict anaerobes) *Methanococcus, Methanobacterium* and *Methanosardna.*

Humus is the organic residue in the soil resulting from decomposition of plant and animal residues in soil, or it is the highly complex organic residual matter in soil. Soil microbes plays important role in the biochemical cycling of elements in the biosphere where the essential elements (C, P, S, N & Iron etc.)

Depending upon the presence or absence of oxygen, non symbiotic N2 fixation prokaryotic organisms may be aerobic heterotrophs (*Azotobacter, Pseudomonas, Achromobacter)* or aerobic autotrophs *(Nostoc, Anabena, Calothrix, BGA)* and anaerobic heterotrophs *(Clostridium, Kelbsiella. Desulfovibrio)* or anaerobic Autotrophs *(Chlorobium, Chromnatium, Rhodospirillum, Meihanobacterium etc).* The organisms involved are *Rhizobium, Bratfyrhizobium* in legumes (aerobic): *Azospirillum* (grasses), Actinonycetes frantic(with *Casuarinas,* Alder).

Several ecofriendly bioformulations of microbial origin are used in agriculture for the effective management of plant diseases, insect pests, weeds etc. eg: *Trichoderma* sp and *Gleocladium* sp are used for biological control of seed and soil borne diseases. Fungal genera *Entomophthora, Beauveria, Metarrhizium* and protozoa *Maltesia grandis. Malameba locustiae* etc are used in the management of insect pests. Nuclear polyhydrosis virus (NPV) is used for the control of *Heliothis* / American boll worm. Bacteria like *Bacillus thuringiensis, Pseudomonas* are used in cotton against Angular leaf spot and boll worms.

Soil receives different toxic chemicals in various forms and causes adverse effects on beneficial soil micro flora / micro fauna, plants, animals and human beings. Various microbes present in soil act as the scavengers of these harmful chemicals in soil. The pesticides/ chemicals reaching the soil are acted upon by several physical, chemical and biological forces exerted by microbes in the soil and they are degraded into non-toxic substances and thereby minimize the damage caused by the pesticides to the ecosystem. For example, bacterial genera like *Pseudomonas, Clostridium, Bacillus, Thiobacillus, Achromobacter etc. and* fungal genera like *Trichoderma, Penicillium, Aspergillus, Rhizopus, and Fusarium* are playing important role in the degradation of the toxic chemicals/pesticides in soil.

□□□

Chapter 2

Soil Chemistry

Chemistry of the Soil is a division of soil science which deals with the study of soil chemical properties, chemical constituents and the reactions in soil.

Soil Chemical properties

There are number of soil chemical properties which affect the plant nutrient availability in soil. The nutrient content and the form depends many soil properties. Very important parameters are soil pH. Electrical conductivity and Cation Exchange Capacity (CEC).

1. Soil pH

It is the negative logarithm of hydrogen ion concentration It can be derived from the dissociation reaction of water.

Dissociation of Water

$$H_2O \rightleftharpoons [OH\text{-}]$$

According to law of mass action, $K_{eq} = \frac{[H^+][OH^-]}{H_2O}$

By substituting weight of One mole of H_2O (i.e., 18 g) and constant Keq. Value (i.e., 5.555×10^{-16}), we get

$$5.555 \times 10^{-16} = \frac{[H+][OH-]}{18}$$

$$[H^+]\,[OH^-] = 5.55 \times 10^{-16} \times 18$$

$$[H^+]\,[OH^-] = 99.99 \times 10^{-16}$$

$$[H^+]\,[OH^-] = 10^{-14}$$

Applying (negative log) on both side and by putting 'p' for negative log, we get,

$$pH + pOH = 14$$

General pH conditions of soil

The soil reaction is measured in terms of soil pH, it ranges from 0 - 14. It is classified into "Slightly acid" (6.0 - 6.5), "Moderately acid" (5.0 - 6.0) and "Strongly acid" (< 5.0) for acidic soil conditions and "Slightly basic" (7.5 – 8.0), "Moderately basic" (8.0 - 9.0) and "Strongly basic" (> 9.0) for alkaline soil conditions. In between 6.5 to 7.5 it is referred as neutral pH.

Causes of soil acidity

In soil, both H^+ and Al^{+3} ions produce acidity. The Al^{+3} produces H^+ ions when it reacts with water (*when pH is below 6: Al^{+3} is the cause of acidity*). Accumulation of soluble acids by carbonic acid (by plant and microbial respiration and atmospheric CO_2) and mineralization of organic matter (produces organic, nitric, sulfuric acids) also causes soil acidity.

Hydrogen ion in any solution or on any particle gives rise to acidic nature of the medium. Let us understand the acidic concept. Molarities and Normality of the solutions are given below:

Molarity of the solution can be prepared by dissolving gram molecular weight of the chemical in given one litre distilled water (volume made is one litre). Summation of atomic weights of the atoms present in the molecule is the molecular weight. Normality of a solution can be prepared by dissolving gram equivalent weight of the chemical in one liter (volume made is one litre). Molecular weight divided by valence gives the equivalent weight. Normality of acid solutions is used to prepare sub levels.

Some Acids Normality is given here: HCL - 12, HNO_3 - 16, CH_3COOH - 18, H_2SO_4 - 36. These have to be used for preparation of lower normality from concentrated form.

2. Electrical Conductivity (EC)

It is also called as total soluble salts in a solution or liquid. It is a measure of total soluble salts in solution, expressed as mmho's/cm or dS/m (or mS/cm).

In irrigation waters, 2 dS/m indicates nearly presence of one gram of salt per liter of water. Similarly in soils, EC of One dS/m indicates presence of one gram of salt per kg of soil.

Preparation of Chemicals

It is necessary to prepare chemicals with different concentrations for conducting reactions during the analysis of various soil chemical parameters. It is required to prepare solutions of different percentages, ppm concentrations, solutions of normality (N) and molarities (M).

Let us understand these preparations: To prepare 1 ppm solution it is required to dissolve 1mg of chemical in 1 liter of distilled water / normal pure water. The ppm means 'one part' in every 'million parts'. That is one liter of water (1000g) is equal to 1,000,000 mg (i.e., million mg). Therefore one mg in 10^6 mg is referred as one ppm. The ppm is also taken for 1kg content in 1 million kg material (i.e., 1000 ton) and also for one gram content in one ton of any material, the concentration remains the same.

Let us prepare 10 ppm of Na using NaCl chemical. That is 10 mg of Na dissolved in one liter of distilled water gives 10 ppm of Na. To take 10 mg of Na how much of NaCl is to be taken, that has to be calculated. In 58.5 mg of NaCl, there is 23 mg of Na. So, in how much NaCl, 10 mg of Na is present ?

$$\frac{58.5\times10}{23}=\frac{585}{23}=25.43\,\text{mg}$$

Let us understand preparations of different concentrations of acids (N and M)

Take an example of preparation of 0.5 N HCl

The concentrated HCl has 12.0 N (to be precise 11.6 N). To prepare one liter of 0.5 N of HCl, it is required to take calculated volume of the acid and add distilled water to make the volume to one liter volume, using $N_1V_1 = N_2V_2$

$$12 \text{ X } V_1 = 0.5 \text{ X } 1000 \text{ ml}$$

$$V_1 = \frac{500}{12} = 41.6$$

That is 41.6 ml of concentrated HCl is taken and diluted with distilled water to one liter gives 0.5 N concentration. The concentration can be standardized by primary alkali standard 0.5 N Sodium bicarbonate ($NaHCO_3$) through titration.

Soil pH importance

- Determines solubility of nutrients
- Before plants can get nutrients, they must be dissolved in soil solution
- Microbial activity also depends on pH

Influence of soil pH on nutrient availability

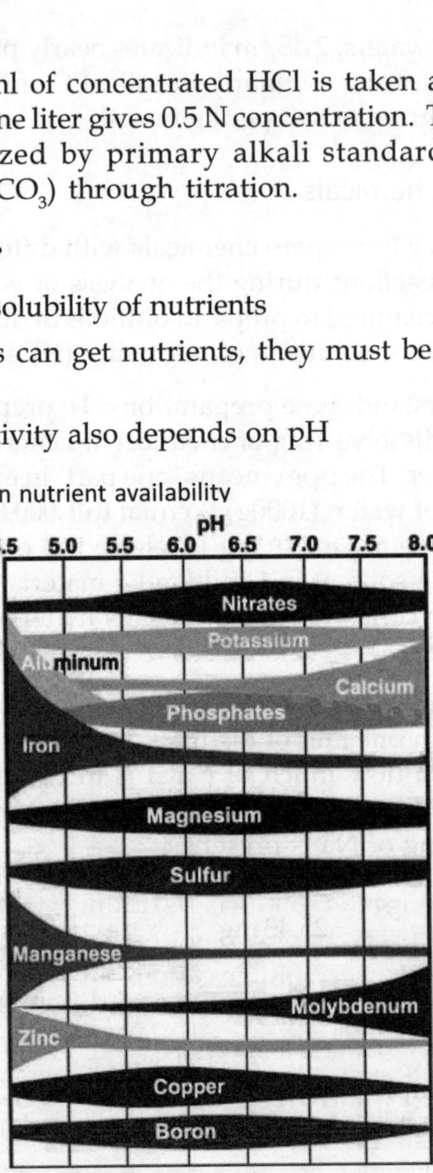

3. Cation Exchange Capacity (CEC)

The soil particles have power to retain cations at the surface of colloids which is referred as the cation exchange capacity. Soil colloids of clay and organic matter have this property due to the presence of negative charges at the surface. The CEC is defined as the sum of cations held by one kilogram of soil. It is expressed in Cmol/Kg soil (older unit used was, milli equivalents per 100 g soil i.e., meq / 100 g soil). Clays like kaolinite have low CEC and pH dependent. Organic matter has higher CEC compared to soils but it is also pH dependent. Montmorillinitic clay has high CEC due to the negative charges developed through loss of cations / isomorphous substitution during formation of these clays. Practices like fertilization, liming, irrigation and addition of organic manures can increase the exchangeable cations.

The relation between CEC and pH

Only 2:1 silicate clays (montmorillonite) do not have pH-dependent CECs. Other type, 1:1 have pH-dependent CEC:

1:1 Kaolinite clays

low pH: low CEC dominated with H^+, Al^{3+} and Fe^{2+}
high pH: high CEC dominated with Ca, Mg K and Na

Oxidic clays

low pH: low CEC dominated with Al^{3+} and Fe^{2+}

Ion Exchange

Ions adsorbed to soil surfaces can be exchanged with ions in soil solution. The cations and anions are attracted to organic colloids and inorganic micelles (clays). These ions come from organic matter/ manures, addition from rain, and weathering of parent materials. Exchangeable cations (on soil surfaces) cannot be removed by leaching, unless there is an exchange reaction between soil solution and exchange site. However, soluble cations (in solution) can be removed easily by leaching. And later there will be reaction between exchange surface and the solution. Then there will be release of some cations into solution. Continues standing and subsequent leaching will make soil defcit of Ca and Mg and dominated by Al, Fe and H.

When soil is dried the exchangeable cations will be held on adsorption sites on soil surfaces. The soluble cations (and anions) precipitate or crystallize as salts. Exchangeable ions on soil surface are the trading places with ions in solution. On soil surfaces, there are: exchangeable and non-exchangeable Ions. Exchangeable cations in outer sphere are weakly held, in contact with soil solution, ready for quick replacement. The cations in *"inner sphere"* are held as non exchangeable, adsorbed by strong bonds or held in inaccessible places (e.g., the K^+ between layers of illite) not as a part of ion exchange.

Diffused double layer theories explain the ion exchange and adsoption. Helmholtz proposed (1879) electrokinetic theory of double layer at a charged surface. The particle surface with negative charges and ions surrounding the particle with positive charges behaves as a condenser. Gouy and Chapman (1910) developed the theory of diffused double layer at a charged surface of soil particles. Asper these, the concentration of counter ions is highest near the charged surface of soil particls and decrease at first rapidly and then asymptotically in solution. Later in a theory given by Stern, which consists of inner layer similar to Helmholtz and the second layer is as that of Gouy and Chapman.

In a clay colloidal with immediate solution containing an electrolyte MA, temperature and pressure being constatnt, the state of equilibrium can be characterized by Donnon's equilibrium:

$$(MA)_l = (MA)_s$$

$$(M)_i\,(A)_i = (M)_o(A)_o$$

Here the ratio being 1, the equilibrium quotient measures the relative adsorption of two ions. The replacibility of an ion goes in the order of : Li<Na<K<Mg<Ca<Ba.

Cation exchange reactions

In many soils, the clay particles carry a net negative surface charge as a result of isomorphous substitution, where structural cations of higher valence are replaced by cations of lower valence (e.g. Si^{4+} is replaced by Al^{3+}). Oxides and SOM also have charged surfaces: with increasing pH (i.e. increasing activity of OH^- ions) H^+ is dissociated from oxide surfaces or from organic functional groups, thus resulting in a negative charge. Dissociation increases with increase in pH and ionic strength, such charge is termed variable charge. At low pH values oxides may bind H+, which results in a positive surface charge. The pH values at which positively charged groups quantitatively equal negatively charged groups (i.e. the net surface charge is zero) are called zero point of charge (ZPC). In soils the overall electro-neutrality is maintained by an excess of electro-statically attracted counter ions in proximity to the charged surface. In case of negatively charged surfaces a diffuse double layer will result where cations are in excess of anions. The excess ions, termed exchangeable cations (or anions) may, by definition, be exchanged with neutral salts. The quantity of exchangeable cations (in meq kg^{-1} dry soil) is defined as the cation exchange capacity (CEC). Similarly the total amount of exchangeable anions is defined as the anion exchange capacity (AEC). In most soils the AEC is quantitatively far less important than the CEC. Exchangeable cations are available to plants, for example through exchange of cations for H^+ liberated by the roots. Exchange reactions are also responsible for the retention of freshly introduced cations into the soil solution. In this way the CEC gives the soil a buffering capacity, which may slow down the leaching of nutrient cations. Since cation exchange is a relatively fast process it modifies the chemical composition of infiltration water if it contains some ions.

Base saturation

Percent of exchange sites occupied by basic cations is reffred as base sturation. Basic cations are those cations other than H^+ and Al^{+3}. Neutral pH (7.0) requires a base saturation of 80% (because most base cations have a + charge of 2).

Rule of ion exchange

The process is Reversible and occur on charge by charge basis which follows Ratio Law that is ratio of exchangeable cations will be same as ratio of solution cations.

Exchangeable Sodium Percentage (ESP)

The extent of sodium on exchange site is measured and expressed as ESP. The percent content of sodium in C.mol (+)/kg soil out of total cations content (i.e., CEC) multiplied by 100 gives the ESP. If it is more than 15, Then the soil particles undergo higher dispersion and become degraded. Hence need to be reclaimed by removing the exchangeable sodium by use of gypsum. One C.mol (+) / kg soil or one milli equivalent of Na require 1.72 tons of gypsum per hectare for replacement.

Soil Chemical reactions

Biogeochemical processes under the soil environment are:

1. Interactions between biota and the atmosphere (e.g. photosynthesis, respiration, N2fixation), and interactions between biota and soil solution (e.g. assimilation and mineralization).
2. Interactions between solution and the soil solid phase (e.g. cation exchange, adsorption, chemical weathering).
3. Chemical reactions in solution (e.g. hydrolysis, complexation reactions) or between solution and atmosphere (e.g. degassing of CO_2)

All these processes modify the chemical composition of infiltration water. For all major solutes the quantitative importance of individual biogeochemical processes can be estimated from input-output budgets and net assimilation rates.

The Adsorption is studied through graphs know as adsorption isotherm. between the amounts of adsorbate (x) adsorbed on the surface of adsorbent (m) with pressure at constant temperature. Different adsorption isotherms have been used to explain : Freundlich, Langmuir and BET theory.

Basic Adsorption Isotherm

In the process of adsorption, adsorbate gets adsorbed on adsorbent.

$$\text{Adsorbate} + \text{Adsorbent} \underset{\text{Aesorption}}{\overset{\text{Adsorption}}{\rightleftharpoons}} \text{Adsorption}$$

$$A + B \rightleftharpoons AB$$

According to Le-Chatelier principle, the direction of equilibrium would shift in that direction where the stress can be relieved. In case of application of excess of pressure to the equilibrium system, the equilibrium will shift in the direction where the number of molecules decreases. Since number of molecules decreases in forward direction, with the increases in pressure, forward direction of equilibrium will be favored.

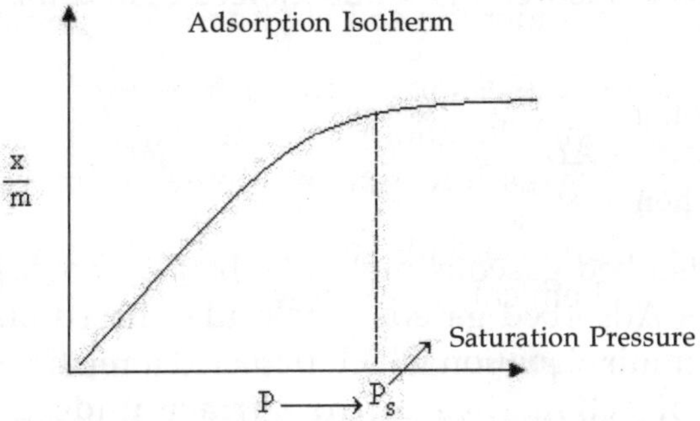

From the graph, we can predict that after saturation pressure Ps, adsorption does not occur anymore. This can be explained by the fact that there are limited numbers of vacancies on the surface of the adsorbent. At high pressure a stage is reached when all the sites are occupied and further increase in pressure does not cause any difference in adsorption process. At high pressure, Adsorption is independent of pressure.

Freundlich Adsorption Isotherm

In 1909, Freundlich gave an empirical expression representing the isothermal variation of adsorption of a quantity of gas adsorbed by unit mass of solid adsorbent with pressure. This equation is known as Freundlich Adsorption Isotherm or Freundlich Adsorption equation or simply Freundlich Isotherm.

$$\frac{x}{m} = kP^{\frac{1}{n}}$$

Where x is the mass of the gas adsorbed on mass m of the adsorbent at pressure p and k, n are constants whose values depend upon adsorbent and gas at particular temperature. Though Freundlich Isotherm correctly established the relationship of adsorption with pressure at lower values, it failed to predict value of adsorption at higher pressure.

Langmuir Adsorption Isotherm

Langmuir proposed an adsorption Isotherm In 1916. This isotherm was based on different assumptions one of which is that dynamic equilibrium exists between adsorbed gaseous molecules and the free gaseous molecules.

$$A(g) + B(S) \underset{\text{desorption}}{\overset{\text{Adsorption}}{\rightleftharpoons}} AB$$

where A(g) is unadsorbed gaseous molecule, B(s) is unoccupied metal surface and AB is Adsorbed gaseous molecule. Based on his theory, he derived Langmuir Equation which depicted a relationship between the number of active sites of the surface undergoing adsorption and pressure.

$$\theta = \frac{KP}{1 + KP}$$

Where è the number of sites of the surface which are covered with gaseous molecule, P represents pressure and K is the equilibrium constant for distribution of adsorbate between the surface and the gas phase. The basic limitation of Langmuir adsorption equation is that it is valid at low pressure only.

At lower pressure, KP is so small, that factor (1+KP) in denominator can almost be ignored. So Langmuir equation reduces to

$$\theta = KP$$

At high pressure KP is so large, that factor (1+KP) in denominator is nearly equal to KP. So Langmuir equation reduces to

$$\theta = \frac{KP}{KP} = 1$$

BET adsorption Isotherm

Brunauer, Emmett and Teller explained that multilayer formation is the physical Adsorption.

One of the basic assumptions of Langmuir Adsorption Isotherm was that adsorption is monolayer in nature. Langmuir adsorption equation is applicable under the conditions of low pressure. Under these conditions, gaseous molecules would possess high thermal energy and high escape velocity. As a result of this less number of gaseous molecules would be available near the surface of adsorbent.

Under the condition of high pressure and low temperature, thermal energy of gaseous molecules decreases and more and more gaseous molecules would be available per unit surface area. Due to this multilayer adsorption would occur. The multilayer formation was explained by BET Theory. The BET equation is given as :

$$V_{total} = \frac{V_{mono} C\left(\frac{P}{P_0}\right)}{\left(1-\frac{P}{P_0}\right)\left(1+C\left(\frac{P}{P_0}\right)-\frac{P}{P_0}\right)}$$

Where Vmono is the adsorbed volume of gas at high pressure conditions.

Reactions involving inorganic carbon

In soil solutions, dissolved inorganic carbon is abundant, and consists of

$H_2CO_3 \rightarrow CO_2(aq) + H_2CO_3$, HCO_3^- and CO_3^{-2}

Degassing of CO_2 is common when soil water emerges. Carbon dioxide which is dissolved in soil water, may react with minerals (including feldspars and calcite) generally resulting in soil solution pH values well above 6. Most natural freshwaters are in this carbonic acid buffer range. The presence of carbonic acid in water gives rise to alkalinity.

Reactions involving organic carbon

Soil organic matter (SOM) can be subdivided into non-humified and humified material. Non-humified substances are slightly altered organic materials after decay of tissue from living organisms. Humified substances are decomposition products include humin, fulvic acid (FA), hymatomelanic acid and humic acid (HA). The concentrations of non-humified organic acids are low The low molecular weight organic acids may play a significant role in mineral weathering. In most soils the contents of HA and FA are considerably higher than those of the non-humified organic acids. The major reason for the importance of FA and HA in soil chemistry is the presence and position of functional groups (particularly carboxyl and phenolic hydroxyl groups), which make FA and HA effective in cation exchange and complexation reactions. Charge characteristics of humic substances depend upon the extent of dissociation of the functional groups. At pH < 3, the HA and FA behave as uncharged polymers, whereas at pH > 3 they behave as negatively charged polyelectrolytes, due to dissociation of carboxyl groups (pH >3 and < 9) and phenolic hydroxyl groups (pH> 9).

Fulvic and humic acids may form metal complexes of high stability through chelation by carboxyl and phenolic hydroxyl groups. Chelation promotes the dissolution of metals from soil minerals. For example, in podzols chelation of Al and Fe in the surface mineral soil and subsequent leaching eventually result in Al and Fe depleted eluvial (E) horizons. The solubility of chelates decreases with increasing sesquioxide content, however, thus resulting in precipitation in the illuvial (podzol B) horizon.

Sulphur Chemistry in Soil

Oxidation of reduced S (e.g. pyrite), and atmospheric deposition of S compounds contributed to sources of sulphate in most natural soils. Atmospheric S (e.g. SO_4^{-2} due to burning of fossil fuels), and of natural origin (e.g. SO_4^{-2} due to volcanic activity. Most S inputs are eventually transfonned into H_2S0_4 and may acidify soil and water. Only S in sea spray is accompanied by base cations and is not acidifying.

The acidifying potential of H_2S0_4 in soils is largely determined by sulphate adsorption. Particularly, Al and Fe oxide surfaces have been reported to be major adsorption sites for soil solution $S0_4^{-2}$.

$$\text{Oxide-OH-O} + S0_4^{-2}\text{-} \rightarrow \text{Oxide-}S0_4\text{-} + \text{OH}$$

where OH- is released and a cation exchange site is created . Alternatively $S0_4$ adsorption may result in the displacement of an $-OH_2^+$ ligand from the oxide:

$$\text{Oxide-}OH_2^+ + S0_4^{-2}\text{-} \rightarrow \text{Oxide-}S0_4\text{-} + H_2O$$

Sulphate adsorption increases with decreasing pH. This pH-dependent adsorption probably results from the amphoteric character of the adsorption sites on oxide surfaces, which become negatively charged upon deprotonation, thus causing anion repulsion. The adsorption of sulphate is negatively affected by the presence of organic matter, even in soils where the contents of Al and Fe oxides are high. This reduction in sulphate adsorption has been hypothesized to be the result of blocking of the sites by organic matter. Sulphate adsorption is generally considered a reversible process and is frequently described by a Langmuir isotherm.

Langmuir and Freundlich adsorption isotherms

Microbiological transformations may play a crucial role in S cycling, as illustrated by the general dominance of organic S forms in soils. Particularly in catchments with significant amounts of wetlands, reduction of $S0_4$ to H_2S and FeS may be an additional important sink for $S0_4$.

Phosphorus Chemistry in Soil

Phosphorus, often one of the essential nutrients, is generally tightly conserved in natural ecosystems. Living organisms contribute significantly to P retention through close coupling of decomposition of organic matter and assimilation. These biological controls on phosphate are dominant in the 0 horizon and the upper mineral soil layers. Inorganic P is adsorbed on clay minerals and on Fe and Al oxides as an innersphere complex through ligand exchange with -OH surface groups. Adsorption processes largely control soil solution P in horizons high in sesquioxides (e.g. B horizons). The bioavailability of P in such horizons depends on the stability of the oxide-P surface complexes and the presence of ligands that can replace P.

Reactions Involving Aluminium

Although Al is one of the most abundant elements in soils, it is not very mobile at the pH values encountered in most soils (i.e. between 6 and 8). The solubility of Al increases in acidic soils (pH <6) and in the presence of complexing ligands. In naturally acidic soils dissolved Al is predominantly in an organically complexed form. Introduction of high loads of mineral acidity results in a significant mobilization of inorganic Al, so that it may become the dominant cation in solution. Dissolved Al is potentially toxic to life and has been hypothesized to be one of the major causes of the occurrence of widespread forest decline. Major sources of dissolved Al as exchangeable Al, organically bound Al, amorphous Al oxide and crystalline Al oxide (e.g. gibbsite). Al solubility in soils is limited and largely based on equilibrium assumptions with respect to a few Al-containing minerals, e.g. gibbsite ($Al_2(OH)_6$).

Chemical Weathering

Together with atmospheric input, chemical and physical weathering of mineral constitutes the major source of base cations in soil. Liberation of base cations through chemical weathering increases the alkalinity of percolating water. The base cation weathering rate and thus the rate of alkalinity production determines the long-term resistance of a soil to acidification (defined as a decrease in pH and base saturation) and is often taken as a measure for the critical load of strong mineral acid to a soil. The critical load is defined here as the maximum strong acid input that can be neutralized without depleting the stores of exchangeable base cations.

□□□

Chapter 3

Soil Fertility

The Soil Fertility: It denotes the status of a soil with respect to the amount and availability of nutrient elements to plants necessary for its growth. Several elements take part in the growth and development of plants, and those absorbed from the soil are generally known as plant nutrients. Besides these uptake, the plant also absorb carbon, oxygen and hydrogen, either from the air or from the water. **In all, 16 elements have been identified and are established as essential for plant growth.**

It is required to understand the soil fertility by considering

i. The nutrient requirement of plants
ii. Supply of nutrients by the soil
iii. Ways in which nutrients are lost from the soil and
iv. Methods by which soil fertility may be maintained or restored

The elements serve as raw materials for growth and development of plants, and formation of fruits and seeds. Most of the essential elements are found in liberal quantities in the mineral soils. In spite of the fact that these are available in plenty, these may not be available to the plants, as they are tied up in mineral and chemical compounds. The roots cannot absorb and deliver them to the growing plants for synthesis, and hence, the need for assessing the plant available amounts of nutrients in the soil and meeting deficiency by application of manures and fertilizers to such soils for optimum crop production.

Inherent mineral source and organic matter content in soil will decide the nutrient availability in soil. Nutrient availability and favourable soil properties decides the soil productivity. All fertile soils may not be fertile but all productive soils are fertile. Soil fertility status will be assessed through soil testing. Soil productivity may be obtained by providing optimum level of different soil properties as required by the plants

Essential Nutrient Elements of Plants

Nutrients are those which support the plant growth and required to attain vegetative and reproductive stages. The Plant Nutrients are: C, H, O, N, P, K, Ca, Mg, S, Cu, Fe, Mn, Zn, B, Mo, Na, Co, Si, V, Cl, and F. Depending on the level in soil they are classified in to **Sufficient** (available in plenty, but not toxic), **Deficient** (when nutrients limits the plant growth due to below critical level) and **Toxic** (sufficiently more to cause toxic injurious effect on plants).

Although plants absorb a large number of elements, all of them are not essential for the growth of crops. The elements are absorbed became they happen to be in the soil solution and those taking active part in the growth and developmental processes are called the essential ones. Some of these are required in large amounts and some in traces. These are classified as major and micro nutrients, and are further classified as follow :

The nutrient elements requiring at comparatively larger quantity for plant growth are called as macro nutrients or major elements (C,H,O, N,P,K, Ca, Mg and S). And those required in smaller quantity are called as micro nutrients (Fe, Zn, Cu, Mn, B, Mo, Cl, Na, V and Si)

There is a requirement of 21 nutrient elements for plants. In periodic table 118 elements are present out of which 21 are identified as nutrient elements. Out of 21 elements 16 are essential (both major and minor elements) and 4 are beneficial (V, Si, Na, Co) (i.e., V - required for symbiotic 'N' fixation, Mo - Nodulation, Si - call wall hardness , Na - involves in water level maintenance in the tissue).

Major nutrients

Primary (Group I): Carbon, hydrogen and oxygen.

Primary (Group II): Nitrogen, phosphorus, potassium

Secondary Nutrients: Calcium, magnesium, sulphur

Micro nutrients: Iron, manganese, boron, zinc, copper, molybdenum and chlorine.

It is found that the presence of some elements though not essential for plant growth but are present as food and feed, are of vital importance to the health of man and animals. The elements within this group are iodine, cobalt and sodium. In addition, there is another group of elements which are toxic to the animals feeding on the plants containing them. These are selenium, lead, thallium, arsenic and fluorine. Elements, such as sodium, fluorine, nickel, lead, arsenic, selenium, aluminium and chromium, when occurring in soils in high available amounts, may also prove toxic to the plants and restrict their growth. Some elements, occurring freely in the soil, are absorbed by the plants as impurities. They may occasionally stimulate growth although they are not essential for plant growth. They include lithium, strontium, tin, radium, beryllium, vanadium, barium, mercury, silver and bromine. Silica is reported to be 'beneficial', particularly in rice but is not classified as essential as per the criteria fixed for this purpose.

1. **Available Nutrient:** Small portion of the total nutrient which is readily absorbable and can be assimilated by plants.
2. **Beneficial Elements:** The element which is not required by plants but enhances / stimulates the growth and yield.
3. **Trace Elements**: Element present in trace quantity (0.0001 ppm).

Deficiency of an element

When an essential element is at a low concentration in the plant tissues, it will result in the decrease in normal growth of the plant, affect the crop yield and produce more or less distinct deficiency symptoms.

- Typical deficiency symptoms are not often clearly defined. Masking effects due to other nutrients, secondary causes like disease, herbicide toxicity or insect infestation can confuse field diagnosis.
- Waterlogged conditions or dry soils and mechanical damage can often create symptoms that mimic deficiencies.
- Deficiency symptoms always indicate severe starvation.

Insufficient levels

When the level of an essential plant nutrient is below the required amount for optimum yields or then there is an imbalance with other nutrients it is considered insufficient. The symptoms of this condition are seldom clearly visible, resulting in poor yield.

Toxicity

Certain essential plant nutrients, if taken up in excess will often cause nutrient imbalances and will result in poor plant growth, delayed maturity, stunted and spindly growth and also show visible symptoms of chlorosis or necrosis.

Arnon and Stout's (1939) criteria of plant nutrient essentiality

i. The plant will be unable to grow normally or complete its life cycle in the absence of particular element.
ii. The element is specific in its role (direct role in the metabolism of the plant) and cannot be replaced by any other element.
iii. The element deficiency can be corrected by supplement of the element only.

Forms of Plant Nutrients

Table 1. Nutrients Essential for plant growth and forms in which taken up by plants Nutrient Chemical symbol Form taken up by plant

Nutrients	Symbol	Forms of absorption
Primary Nutrients		
1. Carbon	C	CO_2, HCO_3
2. Hydrogen	H	H_2O
3. Oxygen	O	H_2O, O_2
4. Nitrogen	N	NH^{4+}, NO^{3-}
5. Phosphorus	P	$H_2PO_4^-$, HPO_4^{-2}
6. Potassium	K	K+
Secondary Nutrients		
7. Calcium	Ca	Ca^{2+}
8. Magnesium	Mg	Mg^{2+}
9. Sulphur	S	SO_4^{2-}

Micro Nutrients		
10. Iron	Fe	Fe^{2+}, Fe^{3+}, chelate
11. Zinc	Zn	Zn^{2+}, $Zn(OH)^2$, chelate
12. Manganese	Mn	Mn^{2+}, chelate
13. Copper	Cu	Cu^{2+}, chelate
14. Boron	B	$B(OH)_3$
15. Molybdenum	Mo	MoO_4^-
16. Chlorine	Cl	Cl^-

Soil Productivity: It is the capacity of a soil in its natural environment to produce crops under a specified system of management and it is expressed in terms of yield. (kg/ha)

The Carbon, hydrogen and oxygen are obtained from air & water. The other thirteen elements are referred to as fertilizer elements and have to be obtained from the soil and or added through chemical fertilizers or organic manures. Their addition in quantities necessary for plant growth will increase the growth rate, dry matter content and yield of the crop.

Plant nutrients are usually absorbed through roots. Roots have the ability to absorb nutrients selectively.

- Root absorption takes place both as active and passive absorption.
- Active absorption takes place as an exchange phenomenon and requires energy.

  Most plant nutrients are absorbed in this manner.
- Passive absorption is part of the transpiration cycle (mass flow). Water and some dissolved solutes are absorbed by this process.
- Gas exchange takes place through the stomata found in leaves. Carbon dioxide required for photosynthesis and oxygen required for plant and respiration are exchanged through the leaves.

The supply of an adequate quantity of a particular nutrient for crop growth depends on both the behaviour of that nutrient in the soil and the ability of the crop root system to utilize it. When these elements are not available to the plant in quantities optimum for growth, the quantity and quality of yield is affected. This requires regular fertilization of crops after determining soil nutrient deficiency.

The plants exhibit hunger signs when they cannot adequately absorb plant nutrients. These symptoms of hunger for nutrients are readily recognizable under field conditions. The hunger can be readily satisfied by the application of fertilizers to the soil.

Hidden hunger

There are no visual symptoms of deficiency but the plant is not producing at its capacity. When the plant reaches the level where symptoms appear, the yield may already have been greatly reduced. Identification of nutrient hunger signs is basic to profitable crop production as it helps in deciding about its application to the soil/ crop. Deficiency symptoms can be categorized into five types.

i. Chlorosis, which is yellowing, either uniform or interveinal of plant leaf tissue due to reduction in the chlorophyll formation.
ii. Necrosis, or death of plant tissue.
iii. Lack of new growth or terminal growth resulting in rosetting.
iv. An accumulation of anthocyanin and/or appearance of a reddish colour.
v. Stunting or reduced growth with either normal or dark green colour or yellowing.

The **Functions of Plant Nutrients,** symptoms of nutrient-wise deficiency and toxicity are described below.

Function of Nutrients

In the nutrition of a crop various nutrients perform distinct functions. Their relative essential role depends upon whether they enter into chemical composition or regulate the various physiological processes in the plant.

Carbon

The plant absorbs carbon dioxide directly from the atmosphere. This combines with water in the presence of light and forms the primary sugars, such as glucose and fructose (fruit sugar). Chlorophyll is the pigment which absorbs the radiant energy of the sun and brings about complex chemical syntheses of carbon dioxide and water resulting in simple sugars. This process is called photosynthesis or synthesis in light. Thus, by the combination of carbon with water, sun's energy is stored in the plant body, and the first carbohydrates are formed in the plant. From the carbohydrates complex sugars, starches, hemicelluloses and celluloses are formed.

These simple sugars also polymerise (chemically combine) into oils and fats. For instance, the soluble carbohydrates (sugars) decrease from 37.5 to 4.5 per cent during the ripening of sunflower seed. Same has been observed about oil in niger seeds.

Oxygen

Oxygen is part of water as well as carbon dioxide. When water combines with carbon dioxide oxygen is evolved:

$$6CO_2 + 6H_2O - C_6H_{12}O_6 + 6O_2$$

Thus, the oxygen evolved in the process equals the volume of carbon dioxide absorbed by the plants. This evolution of oxygen takes place in the process of photosynthesis. The process is reverse in respiration when a simple or a complex sugar, fat or oils breaks up. It requires oxygen and gives out CO_2.

$$C_6H_{12}O_6 + 6O_2 - 6CO_2 + 6H_2O$$

(Glucose)

Six molecules of carbon in the glucose combine with six molecules of oxygen to form six molecules of carbon dioxide. In the process six molecules of water are formed. Thus, oxygen plays a dominant role in the processes of photosynthesis and respiration in plants.

Hydrogen

It is one of the most important elements in the nature. It readily combines with oxygen to form water and with carbon to form complex chemical organic compounds. The growth of plants would only take place if adequate quantity of water is supplied to meet the needs of hydrogen for synthesis of organic substances. When organic compounds either break up in the plant or decompose in the soil or atmosphere, the released hydrogen always combines with oxygen and forms water. Thus, the exchange of hydrogen takes place in either of the synthesis or decomposition (including respiration) processes.

Nitrogen (N): It is one of the major nutrients required in large amounts by plants.

Essentiality

It is not only an essential part of carbohydrates, fats and oils but also an essential ingredient of proteins. Other constituents of proteins are oxygen, hydrogen and nitrogen, usually sulphur and sometimes phosphorus. When proteins decompose through hydrolysis they give out amino acids; reversely, when proteins are formed or synthesized the basic substances are amino acids. Nitrogen is the basic nutrient and makes up 1-4% of dry weight of plants

- It is the primary growth promoter of the plant
- It is a component of chlorophyll

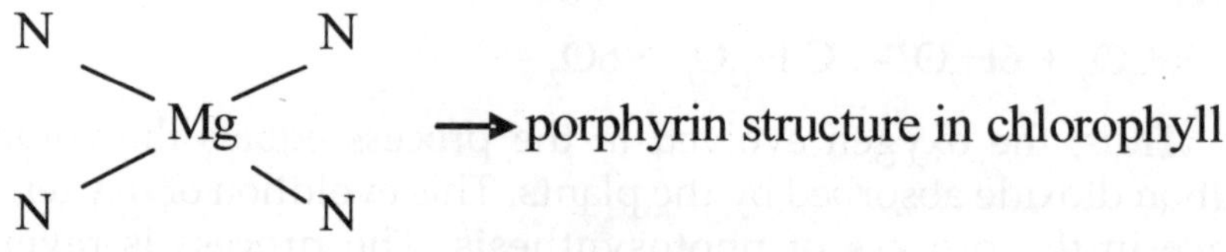

- It is involved in cell wall formation.
- It is the constituent of proteins which are made up of amino acids. Essential amino acids: threonine, valine, tyrosine, isoleucine, leucine, lysine, metheonine, phenyl alanine and Partially essential: histidine, arginine
- It is the constituent of DNA (Deoxyrebo Nucleic Acid : adenine, thymine, guanine and cytosine) and RNA (Rebo Nucleic Acid: adenine, uracil, guanine and cytosine).
- It is involved in carbohydrate synthesis
- When the plant takes up large quantities of nitrogen from the soil the colour of the plant changes to dark-green, indicating the increase of chlorophyll in the plant.
- Since the amount of chlorophyll in the plant determines the carbohydrate synthesis, nitrogen, in a way, may be said to control this activity.
- When there is less uptake of nitrogen, the leaves remain small and pale-yellow in colour.
- As the level of nitrogen supply increases, compared with other nutrients, the extra protein produced enlarges the leaves which provides larger leaf surface for photosynthesis and makes the

leaves more succulent and less coarse, increases the length of the growing season and delays maturity.

- In relation to shoot growth, the root growth is depressed.
- It helps in seed formation and increases the food and feed value of crops. But when the crop plants become more succulent due to larger availability of nitrogen they become susceptible to pests and diseases.

Deficiency Symptoms of Nitrogen in Plants are

The nitrogen-deficient plants are light green in colour. **The lower leaves turn yellow and in some crops they quickly start drying up as if suffering from shortage of water.** The growth is stunted and stems or shoots are dwarfed. In corn if nitrogen deficiency persists the yellowing will follow up the leaf midrib in the typical V-shaped pattern with the leaf margins remaining green. The drying up of lower leaves is generally referred to as firing. The stems are purplish-green. Plants produce fewer lateral branches, reduced number of fruiting branches, and very much reduced number of flowers and bolls. In legumes the growth is stunted and the lower leaves are pale-yellow or brownish in colour. The old, mature leaves are discoloured from base to tip. Under prolonged deficiency twigs become hard and slender. The stems are slender, fibrous and hard.

- Yellowing of leaves because chlorophyll is affected
- General stunted growth
- No deficiency is first observed on older leaves because of its mobile nature
- It shortens blossoms
- Reduces fertility
- It decreases the synthesis of cytokinin (antiaging)
- Alter protein synthesis ; plants become hard

If N is excess

- There will be succulence in plants (softness increases), hence prone to attacked by pest and diseases

- Vegetative growth increases, reproductive growth postponed/ delayed
- Oversize in fruits; dropping of fruits early (pre harvesting stage) stage of crop.
- P and K nutrition is reduced

Phosphorous (P): It is second most essential nutrient required by plants.

Essentiality

It is a constituent of the cell nucleus, essential for cell division and the development of meristematic tissues at the growing points. It makes 0.1 to 0.5% of dry weight of the plant. Therefore, plants which cannot absorb adequate quantities of phosphorus from the soil have small root system and leaves, and their growth is stunted.

- It is involved in N uptake
- Component of phospholipids and ATP
- Important part and parcel of cell membrane
- Enhance seed germination
- Root growth improvement
- It is more deposited in flowers and seeds compared to stem and leaves
- Invertase enzyme is constituted by / controlled by P
- Helps in conversion of sugar to starch
- Important part and parcel of nuclear acid
- In cereals tillering is reduced and maturity is delayed.
- **Phosphorus is particularly helpful in the production of legumes, as it increases the activity of nodular bacteria which fix nitrogen in the soil.**
- It aids the formation of seeds and fruits, particularly in the legumes.
- It stimulates early root growth and development. Optimum quantity of phosphorus available to the crop in combination with nitrogen balances their shoot and root growth.

Deficiency Symptoms of Phosphorous in Plants are

- Generally the plant is dark-green but the lower leaves may turn yellow and dry up. Root and stem growth is stunted and leaves become smaller in size. In corn, leaves and stems have a tendency to become purplish; young plants are stunted and dark-green in colour.
- In deciduous trees the young leaves have dark-green colour while mature ones have bronze or ochre dark-green colour. The new twigs are slender. In many plants the under surface of leaves develops reddish-purple colour. The stems are slender and woody. They bear small, dark-green leaves.
- Root growth is first affected
- Anthocyanine pigments are increased and give red purplish colour to leaves. Eg: maize
- Woody stem formation
- In nature frost resistance is reduced
- Inflorescences becomes small
- Khaira – P deficiency symptoms in paddy (broom stick structure)

If P is excess

- Imbalance in N and K
- Zn and Fe nutrition is reduced along with Ca and Mg
- Zn, Fe, Ca, Mg, farm complex and form insoluble compounds in soil.

Potassium (K): It is the third most essential nutrient required by plants.

Essentiality

Unlike nitrogen and phosphorus, potassium is not a constituent of the carbohydrates, oils, fats and proteins, the substances which form the fabric of the plants.

- Regulation of membrane permeability of plants
- General resistance diseases in plants
- Activates enzymes and photosynthesis

  $C_6H_{12}O_6 \rightarrow CO_2 + H_2O + \text{energy}$

- Present in cell sap and cytoplasm
- It enhances cell division
- Sugars are synthesized in presence of K
- It increases ATP synthesis in Mitochondria
- Cell wall thickness is increases by the potash
- Reduces the rate of respiration
- Hence lead to sugar accumulation
- Helps in translocation of food materials
- But it plays a vital role in the formation or synthesis of amino acids and proteins from ammonium ions which are absorbed from the soil. It is also considered essential in the photosynthetic activity of the leaves.
- When potassium is in short supply the carbon dioxide is synthesized into sugars more slowly than when it is available in optimum quantity.
- The relative concentration of sodium and calcium also influences the activity of potassium in the plant. It helps in moving manufactured food, viz., carbohydrates (sugars) and proteins (amino acids), from leaves to roots.
- It favours the growth of legumes in competition with other plants.
- **The stalks and stems (of plants) are more stiff when an adequate supply is available than otherwise. In consequence the lodging in cereals is reduced.**
- **It increases the plumpness of the grains.**
- **In general it imparts vigour and resistance to diseases.** Some crops, such as potato, tomato, clovers, Lucerne and beans, are more responsive to potassium than other crops. As larger quantities of carbohydrates and proteins are stored, it increases winter hardiness of some plants, such as Lucerne.
- It constitutes 0.8to3.0% of dry matter in cereals.

Deficiency Symptoms of K

The margins of leaves turn brownish and dry up. The stem remains slender. The leaflets become crinkled and curve downward.

The older leaves become yellowish, develop a brown or bronze colour, starting from the tip and edge and gradually affecting the entire leaf, and finally die. In legumes the first symptoms consist of yellow mottling around the edges of the leaf. This area soon dries up and dies. The plants have stunted growth.

- Dieback, leaf scorching, marginal chlorosis (fencing type) Eg: maize leaf
- Rosette in beet, celery, carrot
- Rotting of roots
- Vascular bundle in plant system is affected (smaller) transition becomes difficult
- Fruit/nut drop is seen on deficiency. Eg: Arecanut

If K is excess

- Luxury consumption: It accumulates in the plant part in more amount.

Secondary Nutrients (Ca. Mg and S)

They are normally not supplied as nutrients through commercial fertilizers. They are supplied as amendments to improve deteriorated soil properties, such as pH (acidity/alkalinity), soil structure and ESP (Sodicity).

In acid soils, the pH will be < 6.5 which require lime ($CaCO_3$) for correction to bring the soil pH to near neutral. In alkali the pH will be > 8.3 which require gypsum ($CaSO_4 . 2H_2O$) to bring the soil pH to near neutral. The $MgSO_4$ is applied to soil and sprayed to plant when it is deficient.

Calcium (Ca)

Ca is the secondary nutrient element absorbed by the plant as calcium ion (Ca^{2+}). Calcium sufficiency rate is between 0.2 – 1.0%

Essentiality was identified by *Salm Horstmar* in 1856.

Essentiality

Calcium is essential for the formation of cell-walls, as calcium pectate forms part of the middle layer of the cell-wall. The middle lamella regulates the entry of only those nutrients which are not toxic to the plant. In root-tips calcium is very essential for the meristematic activity or formation of new tissues.

- Stability of cell membrane.
- Ca salt 'lecithin' is a Ca bound lipid involved in cell membrane
- Component of cell wall
- Neutralizes organic acids
- Prevent lodging of plants
- It has very much important role in vegetative growth
- It is required in cell division and chromosome stability
- Genetic makeup is affected
- Ca is required for cell elongation and construction
- Ca is found in meristamatic tissues (growing tips)
- It is involved in root growth and root hairs
- It is required for mitosis and miotic spindle organization
- Ca gives mechanical strength to plant through cell growth
- It stabilize pectins of middle lamella forming Ca pectate
- Involved nitrate assimilation ($NO_2^- \rightarrow NO_3^-$)
- It is required for mitochondria (CHO is broken → starch to get energy)
- It is required for amylase and ATP ase and also kin are enzymes
- Also required for cation - anion balance
- Ca is present in vacuoles and act as counter ion ($Ca^{2+} \rightarrow Mg^{2+}$)
- It also helps to keep up sustained activity of the nodule bacteria in legumes.
- Besides its direct nutrient value, calcium when applied to acid soils increases the
- availability of other nutrients, like phosphorus, nitrogen and molybdenum.
- Excess of calcium in the calcareous soils depresses the uptake of potassium and magnesium. These are secondary effects of calcium on plant growth.

Ca deficiency

Generally the deficiency symptoms due to calcium starvation are localized in new leaves and in bud leaves of plants. In severe cases the terminal bud dies. In tobacco the young leaves making up the terminal bud first become typically hooked and dieback at tips and margins. The stalk finally dies back. In maize the tips of the unfolding leaves gelatinize and when they dry they stick together. In potato a light green band appears along the margins of the young leaves of the bud. The leaves often have a wrinkled appearance. In cotton calcium deficiency makes the petioles bend and later collapse. In vegetables the stems grow thick and woody, and the new leaves are chlorotic. The new growth lacks turgidity. In legumes the nodules developed are small and fewer in number. In citrus the green colour fades along the edges of the leaf and this spreads to areas between veins. The symptoms appear first in immature leaves of deciduous trees, especially those at the top which dieback from tips and margins or along the midribs.

- Dieback from tip of plant
- It cause abnormal mitosis
- The chloroplast become varied size
- It is not mobile in plants however because it is involved in meristanic tissue, young shoots are first affected
- Leaves appear dark in colour. Because of phenol synthesis which oxidises to chitin

  Chitin + amino acid – – – – –→ melanin
- Black heart of cabbage
- Blossom end rot in tomato
- Tip hooking in cauliflower and beet
- Bitter pit of apple

Ca Toxicity

- Plants become succulent - pests and diseases increases
- Affects P, Fe, Mg nutrition

  Eg: $Ca^{2+} + PO^{4=} \rightarrow Ca_3\,(PO_4)_2$ rock phosphate
- It reduces Fe availability

  Magnesium (Mg): Its essentiality was identified during 1891

Essentiality

The chlorophyll development is much reduced when magnesium uptake is restricted because it is an integral part of the pigment.

- It maintains the dark-green colour of leaves and regulates the uptake of other materials, particularly nitrogen and phosphorus.
- It appears to play an important role in the transport of phosphorus, particularly into the seeds.
- It is also said to promote formation of oils and fats, possibly by increasing photosynthetic activity in the leaves.
- Constituent of chlorophyll and chromosomes
- Essential for carbohydrate metabolism
- Required by many crops. Eg: rubber. Coconut, cotto
- Has direct role in photosynthesis
- Only 15-20% of Mg found in plants which is associated with chlorophyll, hence 70 % of Mg is mobile in plants
- It acts as activator of enzymes
- Involved in phosphorylation (energy transfer process)
- Involved in regulation of cellular pH
- Directly involved in transpiration
- It is associated with organic acids like malate and citrate
- It is essential for binding of blossoms (protein synthesis)
- Helps in transport of fats and other nutrients from cell to cell
- Has chelae like link → if Mg increases helps in P uptake
- Required for stabilization of cell membrane

Mg Deficiency: Leads to following symptoms in plants and involve in some functions.

The symptoms of magnesium deficiency at first manifest themselves in old leaves. In tobacco the lower leaves are chlorotic but do not show dead spots. The tips and margins of the leaf are turned or cupped upwards. The stalks are slender. In maize leaves a slight yellow streak develops between the parallel veins in the leaves. In acute deficiency these streaked tissues may dry up and die. The chlorosis in legumes begins at the tip and margins of the lowermost leaf, and progresses between the veins towards the centre of the leaflet. Eventually the tissue between the veins is filled with brown, dead areas.

- Cotton reddening
- Bronzing of citrus
- Interveinal chlorosis in all plants
- Sand drawn disease in tobacco
- Wooly flesh in tomato
- Affects chloroplast size, structure and function
- Protein synthesis is affected/ceased
- As CHO degradation takes place starch accumulation can be seen
- Mg deficiency is observed in older leaves

Mg Toxicity

- Death of roots and leaves
- Roots become brown, cessation of growth
- If Mg is more Ca is affected because of counterion effect
- Uptake of P is affected

Sulphur (S): Sulphur is absorbed by plant as sulphate ion. The sufficiency renge in plant is 0.1-0.4%. Its essentiality is developed by *W. H. Peterson* in 1911.

Essentiality

- Sulphur is found to increase oil content and Responsible for pungency in onion
- Promotes root growth and nodulation
- Most of the S is present in leaves compared to stem

- Involved in photosynthesis i.e., it has role in structural formation of chlorophyll
- S is essential for chloroplast synthesis and S has role in protein synthesis

  Eg: cystine, cysteine, methionine
- It is important in S proteins (amino acids; cysteine and methinoine)
- S is a component of sulpholipids (R – COO – R esters)
- S is constituent of biological membrane and S is involved in lignin and sterol synthesis
- S is involved in N – use efficiency
- Sulphur is a constituent of many proteins, and aids in the formation of chlorophyll and root growth. Plants having an abundant supply of sulphur develop dark-green leaves and extensive root system. In legumes the nodular activity is appreciably increased by adequate supply of sulphur. Due to larger availability of proteins the plant growth is vigorous. It improves the starch control of tubers.

Deficiency Symptoms

Generally the terminal bud remains alive. The chlorosis of younger leaves takes place. The stems harden and sometimes become abnormally elongated and spindly. In legumes the younger leaves turn pale-green to yellow, while terminal buds remain alive.

- Tea yellow diseases
- Upward curling
- Young leaves and tips become dry and die
- Chlorosis occurs unlike N but on youngest parts
- Chlorophyll content in S deficit condition is 40% less than normal condition
- Cell division is related and fruit development suppressed
- S is accompanied by NO^-_3 accumulation in plant cells
- Levels of amino acids is reduced due to S deficiency

S Toxicity

- Scarching of leaves
- Over growth of meristematic tissues

Micronutrients

These are essential plant nutrients required in minute quantity. They are Fe, Mn, Zn, Cu, Mo, B. others are trace and beneficial elements: Na, V, Co, Si, Cl.

Zinc (Zn): The Zn sufficient range in plant is 27-150ppm. Its essentiality of Zn is shown in 1914 by *Maze.*

Essentiality

In a general way zinc is associated with the development of chlorophyll in leaves and a high content of zinc is correlated with a high amount of chlorophyll. In its absence growth is less, buds fall off and seed development is limited. In peach and apricots zinc deficiency symptoms are manifested in leaf. In small trees bronzing of leaves is mitigated by spraying zinc sulphate on leaves. The citrus mottling of leaves may be frequently due to the deficiency of zinc in the plant.

- It helps in photosynthesis by involving chlorophyll synthesis.
- It is essential for synthesis of IAA and cysteine.
- It is involved in oxidation and reduction function.
- It has role in vegetative growth.
- It is involved in synthesis of tryptophan which is a processor of IAA (Auxin which is present in growing tips).
- It has role in protein synthesis it is a cofactor for RNA production.
- It is involved in DNA synthesis also.
- Zn is required for pollen tube growth.
- It has role in N metabolism; metabolites such as amino acids and amides which contain N, their metabolism is mediated by Zn.
- Zn activates carbonic anhydride enzyme which catalyses carbonic acid (H_2CO_3) into CO_2 and H_2O.
- Zn has role in water regulation so it is involved in osmotic pressure maintenance.

Zn deficiency gives following symptoms

Various plant species show different symptoms of zinc deficiency. They are mottled or chlorotic with spots which rapidly enlarge involving secondary and primary veins in succession. The leaves are thick. They have short internodes. On maize seedlings 'white bud' disease is noticed. Extreme deficiency of zinc manifests in chlorotic conditions and in darkercoloured veins of leaves.

- Rosette of fruit trees
- Drenching of citrus
- Sickle leaf of cocoa
- Little leaf of apple
- Reduced chlorophyll content and number of chloroplasts mocuolization
- Plants contain less starch and accumulation of amino acids and amides are seen
- Early drop of nuts/fruits because of reduced auxin
- Abscission ring is formed due to auxin reduction (eg: Arecanut)
- Little leaf, resetting (specially observed in sapota)
- There will be shortage of inter node length
- On server deficiency dieback can be observed
- RNA and ribosome are reduced, so protein synthesis is affected
- White line of patches may be observed on leaves

Zn Toxicity

- If Zn is high Fe uptakes hindered (antagonism) and gives Fe deficiency symptoms
- If Zn is high P nutrition is affected (it forms insoluble P compounds)
- P and Zn are advised to apply to crops separately

Iron (Fe): It is the first micro nutrient identified as essential to plants. It is taken up as ferrous but (Fe^{2+}) by plants. Its concentration of 100-500 mg/kg plants is regarded as sufficient for optimum crop production. Fe is a transitional metal which exhibits +2 and +3 oxidation status i.e., Fe^{2+} and Fe^{3+}.

Essentiality

Although iron does not enter into the composition of chlorophyll, its deficiency manifests itself in chlorosis, yellowing or whitening of leaves.

- The concentration of iron ions plays and important part in the oxidation process in leaf cells.
- When iron is not taken up in adequate quantity, the growth of plants is less vigorous, and seed and fruit development suffer as a consequence of decreased photosynthetic activity in the leaves.
- Too much liming results in iron deficiency. Sever deficiency results in chlorosis and leaves turn white and eventual leaf loss.
- Fe is constituent of cytochrome
- Catalytic role in enzymatic activity
- Fe is required in ferridoxin production
- Ferridoxin is Fe compound presents in chloroplast
- It is required for light reaction where is e – s are released and accepted. It acts as acceptor of e –s
- Involved in biosynthesis of chlorophyll through enzymes
- It is required for pyrol ring in chlorophyll which corrects iron deficiency in leaves
- It plays similar role to Mg in porphyrin structure of chlorophyll
- Fe^{+} photopopuyrin links and acts as processor of chlorophyll formation
- Fe has role in respiration
- Cytochromes contain Fe involved in conversation of pyuric acid in to CO_2 and H_2O. Here ATP's are released
- Ferridoxin is required for reduction of nitrates and sulphates
- Fe activates enzymes involved in respiration
- Like hemoglobin in human plant needs ferridoxin

Fe Deficiency

The iron-starved plants have short and slender stalk. Their terminal buds remain alive but their new leaves show chlorosis of tissues in between the veins, which themselves remain green. In tobacco the young leaves from the terminal buds show chlorotic appearance. The veins of these leaves remain typically green. In young leaves a slight uniform chlorosis is at first noticed. The margins and veins retain green colour. The leaves become pale-yellow and subsequently white. In vegetable crops the new leaves develop light yellow colour in between the veins. Later on the entire leaf becomes yellow. In legumes the leaves turn yellow with the veins remaining green, and on leaves spots of dead tissues appear, particularly at the margins. These dead tissues, in due course, drop away. In citrus the dying of twigs is common, accompanied by chlorosis of leaf tissues in between the veins.

- Chlorosis of young leaves with varied degree of venial cholorosis there will be low chlorophyll content
- Protein synthesis is affected
- Low content of starch and sugars are observed in leaves however higher accumulation of reducing sugar is observed in meristamatic tissues
- Lime induced chlorosis is a term often used for chlorosis associated with high lime content in soils. Here Fe become unavailable to chloroplasts

Fe Toxicity

- Scorching of leaves with rusted appearance
- Reddish strips can be seen on leaves
- Leaves become dark.

Boron (B)

Essentiality identified science 1857; but during 1910 the scientist *Agulhon* discarded that it is absorbed by plant as boric acid (H_3BO_3). However it can also be absorbed in some of its anionic forms like di – hydrogen borate ($H_2BO_3^-$), monohydrogen borate HBO_3^{-2}), borate (BO_3^-) under high pH condition. The sufficiency range of B is 10 – 200 ppm in plants.

Essentiality

One of the most marked effect of boron deficiency observed is the restricted development of nodules on the roots of legumes. Very little nitrogen is fixed in these nodules. Besides, its deficiency influences the growing points of stems, buds and roots. The tissues carrying the minerals and water from the soil to the leaves are also disorganized in the absence of boron. The leaves become brittle. Many deficiency diseases, such as internal cork of apples, heart rot of sugar-beets, top rot of tobacco and cracked stem of celery, are caused by its deficiency.

- It has role in photosynthesis
- It influences phosphorylation process in photosynthesis
- Helps in transport of products of photosynthesis from leaves to seeds/fruits/nuts/tubers/stem.
- Is involved energy utilization of plant cell
- B enhances rooting cuttings through oxidative process
- Cell development (i.e., cell elongation and division)
- Transcolation of sugars found by photosynthesis from leaves to sick
- Polysacdaride synthesis
- It has role in hormone movement and action
- It has role in vegetative growth
- Cytokinine production is mediated by B
- It is involved in fertilization process
- Growth of pollengrains/pollen tubes is affected by B
- Major B is found in blossom
- Pectin borate complexes are formed which gives free maturity to pollen tubes and promotes pollination
- Fruit set is increased by B
- Involved in carbohydrate metabolism

B deficiency: Shows the following symptoms

The deficiency symptoms of this nutrient are usually localized on nerve or bud leaves of the plant. In old leaves with acute deficiency they show twisted growth. The stalks finally dieback at the terminal bud. The leaves thicken and margins roll upwards. The leaf points and margin of older leaves die prematurely. The tubers, besides being small in size, have a ruptured surface. In cotton the effect is localized to terminal buds which dieback, resulting in multi-branched plant. The new bud leaves and petioles have light colour, are brittle and are often deformed in shape. Rosetting due to short internodes is pronounced at the shoot terminals. The legumes also have resetting at the terminal buds. The buds appear as white or light brown dead tissue. The plants have little flowering. New leaves have water-soaked flecks, which become translucent. The fruits have hard, fumy lumps in the rind. In deciduous trees symptoms appear on terminal tissues of twigs.

- Similar to Ca
- Shortage of food supply, hence nut drop and tip drying
- Sterility and malformation
- Rosetting
- Decrease water absorption
- Dieback and corking in apple
- Hard fruit of citrus

B Toxicity

- Abnormal growth of meristomatic tissues are observed
- Higher phenols i.e., formation of gluconate borate complex damages cell membrane

Molybdenum (Mo): Mo is taken up by plants as molebdate ion (MoO_4^{2-}). The sufficiency range is 0.1 – 2ppm in plants.

Essentiality

The presence of molybdenum is very essential for the fixation of atmospheric nitrogen the roots of legumes by nodule bacteria. In plants it is essential for the nitrate-reducing enzyme, as plants well supplied with ammonium do not need it as an essential element. The deformity of 'whiptail' produced in cauliflower is due to the deficiency of molybdenum.

- Constituent of nitrate reductase and nitrogenise enzyme
- Enhance nitrogen fixation in plants (dicots)
- It has role in formation and viability of pollens
- It helps in development of anthers

Mo deficiency

The deficiency is markedly evident in legumes, particularly in the subterranean clover. Molybdenum-starved plants have yellowish to pale-green colour

- Affects nodulation
- Scald (to burn, to heat) of legumes
- Yellow spot of citrus
- It has affect on N metabolism
- Sometimes reddish/purplish colours may appear on leaves
- It was to plants to tiper the stem

Mo Toxicity

- Leaf scorching
- Curling of leaves and stem

Chlorine (Cl): Its first essentiality was first observed by *Boyer et al.* in 1954. The sufficiency range of Cl is 100-500ppm in plants.

Essentiality

- It is involved in turgor maintenance
- Water relations are maintained in the plant cell by Cl
- Operates opening and closing of stomata
- The pH cell sap is controlled by Cl
- Absorption of Ca, K and Mg is affected
- Splitting up of oxygen requires Cl
- Gives resistance to pest and diseases

Cl deficiency

- General retarted growth is seen
- Wilting symptoms appear in plants

Cl toxicity

- Leaf quality is affected
- Sensitivity to some of crops like tobacco, potato

Manganese (Mn): It is absorbed by plants as Manganese ion (Mn^{+2}). The sufficient range of Mn in plant is 20-300 ppm. Mn is also a transitional metal and is present in plant as Mn^{2+} but easily oxdisable to Mn^{3+} and Mn^{4+}. Establishment of essentiality by many scientists was done in 1897.

Essentiality

Manganese is an essential element and appears to have a role in the formation or synthesis of chlorophyll.

- Manganese, probably in association with iron, is a constituent of some respiratory enzymes and some enzymes responsible for protein synthesis from the amino acids formed in the leaves.
- Due to deficiency of manganese the carbohydrate synthesis is disturbed, resulting in retarded growth, decrease in the content of ash and failure to reproduce.
- The leaves and roots of plants deficient in manganese have much less of sugars than those which can absorb sufficient quantity of manganese.
- Constituent of chlorophyll
- Oxidation and reduction function
- Helps in ion movement
- Its influence is there on uptake and utilization of other nutrients.
- Helps in formation of chlorophyll and chloroplasts
- It is involved in oxidation and reduction reaction
- Also takes part in oxygen splitting

$$CO_2 + H_2O + \frac{1}{2} O_2 \xrightarrow[Mn^{2+}]{Cl^-} C_6H_{12}O_6$$

- It activates IAA
- Induces lignification

Mn Deficiency: Shows following deficiency symptoms in plants

In this case also the symptoms are localized to terminal buds which remain alive, but the bud leaves are chlorotic with veins light or dark-green. In legumes the terminal buds remain alive but leaves become light green or yellow with green veins. Later on dead tissues appear on the leaf.

- Gum packets in citrus
- Interveinal chlorosis
- Grey spee in oats
- Chlorophyll and chloroplasts are first affected
- In young leaves of dicots interveinal chlorosis
- Deficiency occurs due to heavy application of P fertilizers
- If there is K deficiency, Mn^{2+} deficiency is also seen
- White patches may be seen on leaves
- Oil palm – leaf base wilting

Mn Toxicity

- Leaves become darker

Copper (Cu): Essentiality was established in 1930 by *Al Samar*. Cu is absorbed by plants as cupric ion Cu^{2+}. The sufficiency range is 5-30 ppm in plants.

Essentiality

In the chloroplasts of leaves there is an enzyme which is concerned with the oxidation-reduction processes. The presence of copper is essential for this enzyme to function. **Thus, copper plays an important role in the process of photosynthesis**. Deciduous fruit trees affected with chlorosis, resetting and dieback recover quickly on application of copper sulphate to the soil in amounts ranging from 0.5-2 lb. per tree.

- Necessary for synthesis of vitamin A
- Catalytic role in respiration
- Has role in transport of electrons in photosynthesis
- It also induces lignifications
- Give disease resistance to plants
- It enhance fertility and seed setting in flowers
- It stabilizes chlorophyll
- It influences cell permeability

Cu deficiency shows following deficiency symptoms in plants

The terminal buds remain alive but wilting or chlorosis of bud leaves takes place with or without spots of dead tissues. The veins of these leaves remain light or dark-green. In legumes the young leaves wilt with or without chlorosis. In extreme deficiency there may occur excessive leaf shedding.

- Multiple bud formation
- Gum pockets and dieback in citrus
- Fruit crack
- Exanthema of fruit trees
- Reclamation disease
- There will be sterility
- Wilting of young leaves
- Deficiency occurs due to heavy application of P and N
- Seed setting will be less
- Lignifications will be affected
- Sclerenchyma, collenchymas cells have wide lumens

Cu Toxicity

- Scorching and curving of leaves
- Cup shaped leaves formation take place

Sodium (Na): it is reffred from Natrium

- It influences the nutrition of K. Eg: Coconut.

Na Toxicity

- Wilting of plants
- >15% of ESP in soil is toxic to plants

Cobalt (Co)

- Constituent of vitamin B12 enzyme

Silicon (Si)

- It gives diseases resistance to plants (required by paddy)
- Si helps in P nutrition
- Fly ash is rich in Si (about 60%)

Nickel (Ni): the sufficiency range is 0.1-10 ppm

- Ni is associated with N metabolism by way of influencing ureas activity
- It facilitates transport of nutrients to seeds

Nutrient Absorption Forms and their Concentrations in Plants

- N is absorbed in the form of nitrate (NO_3) and ammonia (NH_4). In plants N content is > 1.5% below which it is classified as deficient.
- P – is absorbed in the form of $H_2PO_4^-$.

  P content should be more than 0.15% below which is classified as deficit
- K → is absorbed in the form of K^+ and its content is > 0.4%
- Ca → is absorbed in the form of Ca^{2+} → > 0.25%
- Mg → is absorbed in the form of Mg^{2+} → > 0.15%
- S → is absorbed in the form of SO_4^- → > 0.2%
- Zn → is absorbed in the form of Zn^{2+} > 0.1 ppm

- Fe → is absorbed in the form of Fe^{2+}> 50 ppm
- Mn → is absorbed in the form of Mn^{2+} > 5 ppm
- Cu → is absorbed in the form of Cu^{2+}> 5 ppm
- B → is absorbed in the form of BO_3^- > 25 ppm
- Mo → > is absorbed in the form of MoO_4^- 0.1 ppm
- Cl → is absorbed in the form of Cl^- > 0.1 ppm

If any nutrient below the content said above will create nutritional disorders indicated below :

1. Leaf base wilt in oil palm. K, Mn and Mg deficiency
2. Genetic orange spot on leaves in oil palm they can be corrected by applying P and k
3. Crown choke in coconut
   - Emergence of shorter leaves (symptoms)
   - For its correction we can use borax @ the rate of 50g / plant
4. Root wilt in coconut :Due to deficiency of K

   Important symptom is
   - Ribbing and flaccidity
   - Bending of middle outer leaves towards the trunk

   It can be corrected through INM (Integrated Nutrient Management)
5. Tapering or pencil point disease in coconut and arecanut
   - Due to deficiency of K, water logging, root grub, micronutrients like Zn, B deficiency
   - It is found commonly when paddy land is converted in arecanut because of poor aeration
   - Sometimes twisting of internodes this is called as hudimundige

   It can be corrected by improving aeration by reducing water logging conditions and by applying the nutrients

6. Button shedding and immature nut fall
    - Caused due to irregular manuring and deficiency of B and K
    - It is also caused by deficiency of water of excess water
    - It is found in coconut and arecanut
7. Barren nut in arecanut and coconut
    - This is because of K and B deficiency

    It is severe then we can apply 2kg of K and 200g per plant
8. Nut splitting and stem splitting in arecanut
    - Because of B deficiency and also due to B and K deficiency and excess water.

    It can be controlled by spraying 2g/L of B spray at the seedling stage
9. Yellow leaf disease
    - It may be due to N, Mg, K and Mn deficiency
    - It is a general symptom and can be seen arecanut
    - Here we can see Kernel discoloration
    - Young roots turn black and rot

    It can be controlled by application of fertilizers of deficient nutrient
10. Dieback
    - Death of young branches

    It can be corrected by applying Bordeaux mixture of 0.5% (5g/L) and also liming is done.
11. Bean disorder in coffee
    - It is due to short supply of carbohydrate also under unavailability of low light
12. Cocoa wilt
    - This is also called charelle wilt
    - This is due to K deficiency.

13. Slow decline in black pepper
    - Due to nutritional deficiency and fungal activity
    - Moisture stress also causes the symptoms
    - Nematodes problem in roots

    It may be corrected by controlling nematode and fungal attack and supplying moisture and also by the application of P and K

14. Frost damage: It is due to cold condition in the environment for long time
    - This can be controlled spraying 0.1% H_2SO_4 (1ml/L)
    - By irrigation to control frost
    - By using wind breaks
    - Create smoke cover that avoids frost

Soil Testing and Interpretation

Testing of soil for its properties, it may be estimation or determination of its character in the collected soil sample. Soil sample is a representative sample from a unit area collected from 8 to 10 spots and it is called as composite soil sample.

Objective

1. To know the nutrient status and recommend suitable fertilizers and organic manure to corps
2. To understand the status of soil properties, i.e., physical, chemical and biological properties and to suggest effective remedial measures.
3. To generate data for compilation of soil fertility maps.

Soil testing refers to the analysis of soils for various parameters mainly the fertility status of soils and predicting the nutrient requirement of crops. It also includes testing of soils for other properties like texture, structure, pH, Cation Exchange Capacity, water holding capacity, electrical conductivity and parameters for amelioration of chemically deteriorated soils for recommending soil amendments, such as, gypsum for alkali soils and lime for acid soils.

One of the objectives of soil tests is to test the nutrient status of an areas. This information is important for determining whether the soils could supply adequate nutrients for optimum crop production or not. Fertilizer use could be aimed at economic optimum yield per hectare. The National interest would be to obtain the maximum yield from the area under cultivation while the farmer's interest would be to obtain profitable yields and not necessarily the maximum yields. Indiscriminate use of fertilizer is not an answer to any one of the problems as this not only increases the cost of crop production but also results in deleterious effects on soil fertility. The concept of balanced nutrition of crops also guides the use of plant nutrients in a definite proportion as required by the crops which is possible only by soil test.

It helps in understanding the inherent fertility status of the soils. Further, various factors also be responsible for poor crop production but soil fertility status assumes a greater importance. Each fertilizer recommendation based on a soil analysis should take into account the soil test value obtained by the accurate soil analysis, the research work conducted on a crop response to fertilizer application in a particular area and the practices and level of management of the concerned farmer. To have better results of soil test, one should have correct method of soil sampling.

Soil Sampling

The method and procedure for obtaining soil samples vary according to the purpose of sampling. Soil sampling for agricultural purpose is done for soil fertility evaluation and fertilizer recommendations for crops. The results of soil analysis will be good if the soil sample correct. Thus, the efficiency of soil testing service depends upon the care and skill with which soil samples are collected. It is to be noted that the most mportant phase of soil analysis is accomplished not in a laboratory but in the field where soils are sampled. Soils vary from place to place. In view of this, efforts should be made to take the samples in such a way that it is fully representative of the field. Only few grams to 20 gram of soil is used for each determination and represents the entire surface 0-22 cm of soil, weighing about 2 million kg/ha.

Sampling procedure

A slice of the plough-layer is cut at intervals of 15 to 20 steps or according to the area to be covered. Generally 10 to 20 spots must be taken for one composite sample depending on the size of the field.

Collect surface litter sperately; obtain a uniformly thick slice of soil from the surface to the plough depth from each place. A V-shaped cut is made with a spade to remove one inch slice of soil. The sample may be collected and put in a clean bucket. In case of hard soil, samples are taken with the help of augur from the plough depth and collected in the bucket.

Pour the soil from the bucket on a piece of clean paper or cloth or plastic sheet and mix thoroughly. Spread the soil evenly and divide it into 4 quarters. Reject two opposite quarters and mix the rest of the soil again. Repeat the process till left with about half kg of the soil, collect it and put in a clean cloth / plastic bag. Each bag should be properly marked to identify the sample. The bag used for sampling must always be clean and free from any contamination. If the same bag is to be used for second time, turn it inside out and remove the soil particles and reuse. Write the details of the sample in the information sheet including Farmers name, Survey number, village, Gram Panchayth, Taluk / block, and district. Put a copy of the information sheet in the bag and keep one copy with you. Tie the mouth of the bag carefully.

The sheet accompanying the sample must include the information on nature of soil, hardness and permeability of soil, salinity cause and source, if known, relief, seasonal rainfall, irrigation and frequency of water logging, water table, soil management history, crop species and conditions of plant cover and depth of the hard pan or concretion if present.

Nitrate, nitrite and ammonium determinations must be carried out on samples brought straight from the field. These samples should not be dried. However, the results are expressed on oven dry basis by separately estimating moisture content in the samples.

After grinding, the soil is screened through a 2 mm sieve. If the soil is to be analyzed for trace elements, containers made of these metals must be avoided during grinding and handling. Sieves of different sizes can be obtained in stainless steel. Aluminium or plastic sieves are useful alternative for general purposes.

Precautions

- Do not sample unusual area like unevenly fertilized, marshy, street path, old channel, old bunds, area near the tree, site of previous compost piles and other unrepresentative sites. It should not be collected all along the borders. It is required to collect 4-5 kg soil sample to get half kg composite soil sample.

- For a soft and moist soil, the tube auger or spade is considered satisfactory. For harder soil, a screw auger may be more convenient. It is required to collect the soil sample in zig - zag pattern, because of it is collected in a straight line, there will be chances of some errors. It should not be collected below the trees because there will be shade and root effect. Root effect is called as "Allelopathic Effects". Land has to be divided before sampling uniformly w.r.t. soil colour, texture, slope, vegetation etc.
- Where crops have been planted in rows, collect samples from the middle of the rows so as to avoid the area where fertilizer has been band placed. It should not be collected where manure is dumped previously.
- Avoid any type of contamination at all stages. Soil samples should never be kept in the store along with fertilizer materials and detergents. Contamination is likely when the soil samples are spread out to dry in the vicinity of stored fertilizers or on floor where fertilizers were stored previously.
- Before putting soil samples in bags, they should be examined for cleanliness as well as for strength. It is essential to have equipment and bags clean. It should not collected in fertilizer bags.
- Information sheet should be clearly written with copying pencil.

Time of Sampling

Dry days or Summer is best for soil sampling i.e., April and May months in India. Because there will be less moisture and depletion of nutrients. However, at any point of time sampling may be done depending upon the need of the hour.

Soil Sampling for Land Capability Classification

Sampling of soil is done for nutrient status or reclamation or evaluation to specific purpose. Evaluation of specific purpose means to classify the entire land based on its capability. There are 8 classes based on certain properties.

Class I – IV – Suitable for Agriculture

Class V – VIII – Suitable for Forest

Soil Sample Collection in Plantation Crops is similar to Forest Plants

1. In coconut we have to collect the soil sample of 2 – 4 m radius, depth 0 – 30, 30 – 60cm
2. Arecanut: 0.75m radius, 0 – 30 cm depth
3. Coffee: at a distance of 0.45m, in coffees roots are in deep depth, so we take 3 samples i.e., <30cm, 30 – 45cm and 45 – 90 cm.
4. Tea: distance 0.45m, depth 0 – 22.5 and 22.5 – 37.5cm
5. Oil palm: distance 1.5m, depth 0 – 30cm
6. Cashew:2m, depth 0 – 30cm
7. Beetle wine: 0.45m, depth 0 – 30cm
8. Black pepper: 0.45m, depth 0 – 30cm
9. Clove: 0.45m, depth 0 – 50cm
10. Cinnamon: 0.45m, depth 0 – 50cm
11. Nut mug: 0.45m, depth 0 – 50cm
12. All Spices: 0.45m, depth 0 – 50cm
13. Rubber: 1.0m, depth 0 – 50cm

Soil Testing Parameters

I. Soil Physical properties

1. Texture: (sand, silt, clay)
2. Coarse textured soil – Deficiency of N, K, B, Zn, S
3. Moisture: Soil Moisture Content (SMC)
4. Particle density
5. Moisture Holding Capacity (MHC)
6. WHC (Water Holding Capacity): If WHC of soil is more that indicates the soil is finer.
7. Field Capacity
8. Bulk density Lower bulk density values are good for aeration to plants

II. Soil Chemical Properties

1. Soil pH:

   - Acidic soil (<6.0) – deficiency of P, B, Mo, N
   - If it is alkaline (>8.0) deficiency of P i.e., in the form of $Ca_3(PO_4)_2$, N i.e., (NH_3) and metallic elements – they becomes unavailable
   - In neutral range almost all will be available in plenty to plants

   The soil pH is the negative logarithm of the active hydrogen ion (H+) conc. in the soil solution. It is the measure of soil sodicity, acidity or neutrality. It is a simple but very important estimation for soils, since soil pH influences to a great extent the availability of nutrients to crops. It also affects microbial population in soils. Most nutrient elements are available in the pH range of 5.5 to 6.5.

2. EC: Higher in the salt content, greater is the tension with which water is held in soil.

Soil Electrical Conductivity(EC)

The soils having pH value more than 8.0-8.5 may have the following special features:

- Presence of excessive amounts of soluble salts.
- Presence of excessive amounts of sodium on the exchange complex.

The electrical conductivity (EC) is a measure of the ionic transport in a solution between the anode and cathode. This means, the EC is normally considered to be a measurement of the dissolved salts in a solution. Like a metallic conductor, they obey Ohm's law.

Since the EC depends on the number of ions in the solution, it is important to know the soil/water ratio used. The EC of a soil is conventionally based on the measurement of the EC in the soil solution extract from a saturated soil paste, as it has been found that the ratio of the soil solution in saturated soil paste is approximately two-three times higher than that at field capacity. As the determination of EC of soil solution from a saturated soil paste is cumbersome and demands 400-500 g soil sample for the determination, a less complex method is normally used. Generally a 1:2 soil/water suspension is used.

3. Organic Carbon (from Organic Matter in soil)

   The organic matter in soil has several functions of organic matter in soil.

- Supplies N,P, S and other secondary and micro-nutrients for plant growth.
- Increases CEC in a soil.
- Have large surface area and has high CEC.
- Holds up to twenty times their weight of water.
- Holds cations and anions and releases them slowly.
- The ratio of C:N:P:S is 10:1:.5:0.1.
- Effects the breakdown of pesticides and weedicides.
- Chelates micronutrients such as Zn, Mn and Cu making them more available to plant roots.
- Buffers extreme acidity, salinity and alkalinity.

Organic matter estimation in the soil can be done by different methods. Loss of weight on ignition can be used as a direct measure of the organic matter contained in the soil. It can also be expressed as the content of organic carbon in the soil. It is generally assumed that on an average organic matter contains about 58% organic carbon. Organic matter/organic carbon can also be estimated by volumetric and colorimetric methods. However, the use of potassium dichromate ($K_2Cr_2O_7$) involved in these estimations is considered as a limitation because of its hazardous nature. Soil organic matter content can be used as an index of N availability (potential of a soil to supply N to plants) because the content of N in soil organic matter is relatively constant.

It is done by Volumetric method (Walkley and Black, 1934)

Organic C to total organic matter conversion factor for surface soils vary from 1.724 to 2.0. A value of 1.724 is commonly used, although whenever possible the appropriate factor be determined experimentally for each type of soil.

OM contains proteins $\xrightarrow{\text{converted into}}$ amino acids $\xrightarrow{\text{dehydrated}}$

amines (NH_2) $\xrightarrow{\text{undergoes}}$ Nitrification

0.85% OC means 0.85kg OC in 100kg soil

If 100kg soil has 0.5kg OC it is equal 0.85 kg OM (factor is 1.723)

100kg soil → 0.6kg OC → 1.02kg OM

In one acre, 10^6kg soil $\rightarrow \frac{0.85 \times 10^6}{100} = 8500kg$ OM

10^6kg soil $\rightarrow \frac{1.02 \times 10^6}{100} = 10200kg$ OM

If it is to raise 1% OC it is required to apply 17 tones of OM per acre, continuosly for 4-5 years.

Interpretation

- It is the indicator of organic matter. In any plant sample OC will be 50-60%
- If we estimate OC, it can be converted to OM (organic matter) by multiplying 1.723. OC is used for knowing available N status in the soil
- If OC is < 0.5% - low

  0.5-0.75% - medium

  >0.75% - high

Soil Available Nitrogen

Available N is estimated by $KMnO_4$ method. Whereas OC is estimated by $K_2Cr_2O_7$ method.

Mineralizable Nitrogen (Subbia And Asija, 1956)

In case of soils, mineralizable N (also organic C) is estimated as an index of available nitrogen content and not the total nitrogen content. The easily mineralizable nitrogen is estimated using alkaline KMnO4, which oxidizes and hydrolyses the organic matter present in the soil. The liberated ammonia is condensed and absorbed in boric acid, which is titrated against standard acid.

Available N in soil

<280kg/ha - low

280 - 560kg/ha - medium

>560 kg/ha - high

∴ if a soil status indicates 170 kg N/acre or 425 kg/ha, it is in the medium state.

All the crops are defined with specific recommended dose of fertilizer (RDF) which are given in Package of Practices of farming Universities of different areas.

Eg: Paddy - 75:75:90 kg NPK/ha

As per the soil status it is required to change NPK doses in RDF

Soil fertility low or high ± 5 to 15 kg nutrient in RDF/ acre

Soil fertility more ± 25 % of RDF

Eg: Maize - 150: 75:37.5 kg NPK/ha RDF

187.5:75:28 kg/ha after soil testing for low; medium; high status

N undergoes certain steps in soil

1. **Aminization:** It is one of the first steps in the decomposition of nitrogenous materials. Simple proteins are hydrolyzed to form amines and amino acids.

   Proteins = $R - NH_2 + CO_2$ + energy + other products

2. **Ammonification:** The amines and amino acids released by aminization are used by other soil heterotrophs and broken down further to ammonical compounds. This process is called as ammonification, it can be represented as follows:

   $R - NH^2 + HOH = NH^3 + R + OH$ + energy

   The ammonia produced by this process may be used in several ways in the soil.

- ❒ Be converted to nitrite and nitrate by the process of nitrification
- ❒ Be used directly by plants
- ❒ Be used directly by soil micro organism

- Be tied up by certain types of soil clays.

3. **Nitrification:** The conversion of ammonium nitrate is called nitrification. It is oxidation process and releases energy for the use of soil microbes. The conversion is a two step process in which ammonium is first converted to nitrite (NO_2) and nitrite is converted to nitrate (NO_3). The conversion of ammonium to nitrite is performed by a number of soil microbes but the most important is a group of obligate autotrophic bacteria known as nitrosomonas.

   $2NH_4^+ + 3O_2 = 2NO_2^- + 2H_2O + 4H^+$

   The conversion nitrite is also performed by a number of soil microbes but is performed mainly by another group of obligate autotrophic bacteria known as nitrobacter

   $2NO_2^- + O_2 = 2NO_3^-$

   The resulting NO_3^- is highly mobile in soils and is easily lost from the soil with water that moves downwards and laterally through a soil profile. The $NO_3^{-\text{is}}$ a potential pollutant if it reaches ground water

4. **De - nitrification:** conversion of nitrates into nitrogen. This is very much important for maintenance of atmospheric balance NO_2 is the laughing gas is the intermediate product.

5. **N - Immobilization:** Immobilization of N refers to the absorption of plant available N forms (NH_4^+ and NO_3^-) by microbes and their transformation into amino acids and protein. This form of N is not available for plant uptake and has been "immobilized" within the microbe tissues.

   Whenever there is immobilization there will be N deficiency in plants. This condition is seen when organic matters with wide C:N ration (more carbon to less nitrogen) are applied to soil.

Soil Available Phosphorus

It is estimated with the use of diffrenet extractant under different soil pH conditions i.e., first it is required to estimate the soil pH because,

In Acidic conditions : Fe and A1 dominate

In Alkaline conditions : Ca and Mg dominate

Available Phosphorus depends on soil pH

Two methods are most commonly used for determination of available phosphorus in soils: Bray's Method No.1 for acidic soils and Olsen's Method for neutral and alkaline soils.

In these methods, specific coloured compounds are formed with the addition of appropriate reagents in the solution, the intensity of which is proportionate to the concentration of the element being estimated. The colour intensity is measured spectrophotometrically.

Available Potassium

Potassium present in the soil is extracted with neutral ammonium acetate of 1 molarity. This is considered as plant available K in the soils. It is estimated with the help of flame photometer.

Available Sulphur

Available sulphur in mineral soils occurs mainly as adsorbed SO_4 ions. Phosphate ions (as monacalcium phosphate) are generally preferred for replacement of the adsorbed SO_4 ions. The extraction is also carried out using $CaCl_2$ solution. However, the former is considered to be better for more efficient replacement of SO_4 ions. Use of Ca salts have a distinct advantage over those of Na or K as Ca prevents deflocculation in heavy textured soils and leads to easy filtration. SO_4 in the extract can be estimated turbidimetrically using a spectrophotometer. A major problem arises when the amount of extracted sulphur is too low to be measured. To overcome this problem, seed solution of known S concentration is added to the extract to raise the concentration to easily detectable level through Barium sulphate precipitation method

Determination of exchangeable Calcium and Magnesium

Exchangeable cations are usually determined in a neutral normal ammonium acetate extract of soil. Extraction is carried out by shaking the soil : extractant mixture followed by filtration or centifugation. Calcium and magnesium are determined either by EDTA titration method.

The EDTA titration method developed by Chang and Bray (1951)

A known volume of solution is titrated with standard versenate 0.01N solution using muroxide (ammonium purpurate) indicator in the presence of NaOH solution. The end point is a change of colour from orange red to purple at pH 12 when the whole of calcium forms a complex with EDTA (Ethylene damine-teraacitic acid).

The method is based on the principle that calcium, magnesium and a number of other cations form stable complexes with versenate (ethylendiaminetetraacetic acid disodium salt) at different pH.

Magnesium in solution can be titrated with 0.01N EDTA using Eriochrome black T dye as indicator at pH 10 in the presence of ammonium chloride and ammonium hydroxide buffer. At the end point, colour changes from wine red to blue or green. When calcium is also present in the solution this titration will estimate both calcium and magnesium. Beyond pH 10 magnesium is not bound strongly to Erichrome black T indicator to give a distinct end point.

Micronutrients

For estimation of micronutrients also, it is the plant available form which is critical and not the total content. The major objective of soil test for micronutrients, like macronutrients, is to determine whether a soil can supply adequate micronutrients for optimum crop production or whether nutrient deficiencies are expected in crops grown on such soils. Most commonly studied micronutrients are Zn, Cu, Fe, Mn, B and Mo and the same have been dealt with here.

Chemical characteristics of saline, non-saline sodic and saline sodic soils

Soil	EC (dS/m)	ESP	pH
Saline	> 4.0	< 15	< 8.5
Sodic (non-saline)	< 4.0	> 15	> 8.5
Saline Sodic	> 4.0	> 15	> 8.5

Source: Richards, 1954

The estimation of elements in the extract is done with the help of Atomic Absorption Spectrophotometer (AAS). Critical limits for DTPA (Diethylenetriaminepentaacetic acid) extractable micronutrient elements as proposed by Lindsay and Norvell (1978) are given

Table 2. Critical limitsd for DTPA extractable micronutrients

Availability	Micronutriens (µg/g soil)			
	Zn	Cu	Fe	Mn
Very low	0-0.5	0-0.1	0-2	0-0.5
Low	0-5.1	0.1-0.3	2-4	0.5-1.2
Medium	1-3	0.3-0.8	4-6	1.2-3.5
High	3-5	0.-8-3	6-10	3.5-6
Very high	> 5	> 3	> 10	> 6

Principle of Atomic Absorption Spectrophotometry

The procedure is based on flame absorption rather than flame emission and upon the fact that metal atoms absorb strongly at discrete characteristic wavelengths which coincides with the emission spectralines of a particular element. The liquid sample is atomized. The hollow cathode lamp which precedes the atomiser, emits the spectrum of the metal used to make the cathode. This beam traverse through the flame and is focused on the entrance slit of a monochromator, which is set to read the intensity of the chosen spectral line. Light with this wavelength is absorbed by the metal in the flame and the degree of absorption being the function of the concentration of the metal in the flame, the concentration of the atoms in the dissolved material is determined. For elemental analysis, a working curve or a standard curve is prepared by measuring the signal or absorbance of a series of standards of known concentration of the element under estimation. From such curve, concentration of the element in unknown sample is estimated.

Atomic Absorption Spectroscopy can be successfully applied for estimation of Zn, Cu, Fe and Mn. For specific estimation on AAS, hollow cathode lamps, specific to specific elements are used

Estimation by colorimetric method

The hot water soluble B in the filtered extract is determined by the azomethine-H or curcumin colorimetric method.

Molybdenum (Mo) is a rare element in soils, and is present only in very small Amounts

In the soil solution Mo exists mainly as HMoO4 ion under acidic condition, and as MoO_4^{2-} ion under neutral to alkaline conditions.

Because of the anionic nature of Mo, its anions will not be attracted much by the negatively charged colloids, and therefore, tend to be leached from the soils in humid region. Molybdenum can be toxic due to greater solubility in alkaline soils of the arid and semi-arid regions, and deficient in acid soils of the humid regions. In plants a deficiency of Mo is common at levels of 0.1 μg/g soil or less.

In case of Molybdenum, ammonium acetate and/or ammonium oxalate extraction is usually carried out. Estimations can be done both by AAS and colourimetric methods

Calibration of Soil test with crop response correlation

The soil testing and soil fertility management programmes have been given adequate importance for sustaining crop production and balanced fertilization in Indian Agriculture. Its implementation has gradually progressed and met with varying degree of success, depending upon areas and even according to the agency responsible for implementing the programme. In the initial stages based upon the simple fertilizer trials on farmers' fields and response to fertilizer application, soil fertility status was categorized into low, medium and high classes.

It may be noted that generally the percentage uptake of applied fertilizer nutrient (N,P,K) varies in different situations as follows:-

Nitrogen - 30% to 60%

Phosphorus - 10 to 30%

Potash - 20 to 60%

Micronutrients Zinc - 2 to 5%

The scope of improving the fertilizer use efficiency is enormous. A perfect soil test based fertilizer use recommendation and application of fertilizer nutrients in optimum combination and required doses can enhance the efficiency of fertilizer use and thus the profitability of fertilizer use in crops.

It enables the literate and enlightened farmers to conduct their own on-the-spot analysis and interpretation of the test result without the assistance of an official extension agent. This is very important in India where the ratio of extension worker to farmer population is very low.

Soil Fertility Evaluation Techniques

Basic aim of the soil testing is to assess the fertility status of the soils so as the recommendations for the fertilizer use and soil amendments, if necessary, can be made. There are, however, other techniques which can be employed for soil fertility evaluation. Different techniques can be used alone or in combination with each other but the soil testing is still most commonly employed technique (Dev, 1997).

Different techniques are listed and briefly described below:-

I. Nutrient deficiency symptoms

II. Biological tests

III. Plant analysis

I. Nutrient Deficiency Symptoms

The appearance of a plant has long been a clue to its nourishment or lack of it. When a plant badly needs a certain plant food, it shows starvation signs called deficiency symptoms or hunger signs. These symptoms are nutrient specific and show different patterns in crops for different essential nutrients. It is a good tool to detect deficiencies of nutrients in the field. However, it requires learning to identify nutrient deficiencies. In applying this technique, one must develop diagnostic proficiency through practice and close observation. It may be pointed out that deficiency symptoms in many cases are not always clearly defined and in cases, the symptoms can be common to other causes or may be masked by other nutrients. Some typical examples are given below:

- N deficiency can be confused with S deficiency.
- Calcium deficiency can be confused with B deficiency
- Fe deficiency can be confused with Mn deficiency
- Leaf stripe disease of oat can be confused with Mn deficiency
- Effect of virus, little leaf etc. can be confused with zinc deficiency, B deficiency.
- Brown streak disease of rice can be confused with Zn deficiency.

Deficiency symptoms always indicate severe starvation and therefore the crop may have suffered before the deficiency symptoms appear. Many crops start losing yields well before deficiency signs start showing. This yield limiting condition is called Hidden Hunger which refers to a situation in which a crop needs more of a given nutrient and yet has not shown any deficiency symptom. The nutrient content is above the deficiency symptom zone but still considerably below that needed for optimum crop production. In this case, significant responses can be obtained with application of nutrients even though no recognizable symptoms have appeared. To over-come such conditions, it is always recommended to confirm deficiency problem with other diagnostic techniques and most common is soil testing. Nutrient wise deficiency symptoms are described in detail in

II. Biological Tests

a) **Field Tests**: The field plot technique essentially measures the crop response to nutrients. In this, specific treatments are selected, randomly assigned to an area of land, which is representative of the conditions. Several replications are used to obtain more reliable results and to account for variations in soil. Field experiments are essential in establishing the equation used to provide fertilizers recommendations that will optimise crop yield, maximise profitability and minimise the environmental impact of nutrient use. When large numbers of tests are conducted on soils that are well characterized, recommendations can be extrapolated to other soils with similar characteristics. Field tests are expensive and time-consuming; however, they are valuable tools and are widely used by scientists. They are used in conjunction with laboratory and greenhouse studies in calibration of soil and plant tests.

b) **Greenhouse tests**: The greenhouse techniques utilize small quantities of soil to quantify the nutrient-supplying power of a soil. Generally, soils are collected to represent a wide range of soil chemical and physical properties that contribute to the variation in availability for a specific nutrient. Selected treatments are applied to the soils and a crop is planted that is sensitive to the specific nutrient being evaluated. Crop response to the treatments can then be determined by measuring total plant yield and nutrient content.

c) **Laboratory tests**: Neubauer seedling method - The Neubauer technique is based on the uptake of nutrients by a large number of plants grown on a small amount of soil. The roots thoroughly penetrate the soil, exhausting the available nutrient supply within a short time. The total nutrients removed are quantified and tables are established to give the minimum values of macro and micronutrients available for satisfactory yields of various crops.

d) **Microbiological methods**: In the absence of nutrients, certain micro-organisms exhibit behaviour similar to that of higher plants. For example, growth of Azotobacter or Aspergillus niger reflects nutrient deficiency in the soil. The soil is rated from very deficient to not deficient in the respective elements, depending on the amount of colony growth. In comparison with methods that utilize higher

plants, microbiological methods are rapid, simple and require little space. These quick laboratory tests are not in common use in India.

III. Plant Analysis

This involve two approaches. One is analysis of plant in the laboratory and the other is tissue test on fresh tissue in the field. The basis for plant analysis is that amount of a given nutrient in a plant is directly related to availability of the nutrient in the soil.

Plant analysis is used to:

- Confirm a diagnosis made from visible symptoms;
- Identify hidden hunger where no symptoms appear;
- Locate soil areas where deficiencies of one or more nutrients occur;
- Determine whether applied nutrients have entered the plant;
- Learn about interactions among various nutrients;
- Study the internal functioning of nutrients in plants;
- Suggest additional tests or studies in identifying a crop production problem.

Plant Analysis is generally not done in the soil testing laboratories. Plant analysis is used more for fruit and vegetable crops. Because of the perennial nature of the fruit crops and their extensive root systems, plant analysis is especially suitable for determining their nutrient status. As more is known about plant nutrition and nutrient requirements throughout the season, and as application of nutrients through irrigation systems is possible, plant analysis assumes greater importance. Also, to produce high yields, it is helpful in monitoring the plant through its growing season. It is becoming more useful for row crops and forage crops.

The critical nutrient concentration (CNC) is used in interpreting plant analysis results and diagnosing nutritional problems. CNC is the level of a nutrient below which crop yield or quality is unsatisfactory. In India, a great deal of information has been generated on critical nutrient level in crops especially for micronutrients and sulphur.

Plant Analysis: Ratio of nutrients in plants. i.e., N to P, N to K, P 'to K' is used to know nutrient requirement

The plant analysis has 3 main applications:

- To confirm a suspected nutrient element deficiency when visual symptoms are present
- To monitor the plant nutrient element status in order to determine whether each tested nutrient is in sufficient concentration for optimum yield
- To serve as the basis along with a soil test for making fertilizer recommendation for fruit and nut crops
- The demand for this service should increase in India in future as research emphasizes opportunities to manage nutrient availability during the growing **season.** An important phase of plant analysis is sample collection. Plant composition varies with age, the portion of the plant sampled, the condition of the plant, the variety, the weather and other factors. Therefore, it is necessary to follow proper sampling instructions. For plant analysis, a typical plant part is selected which indicates the nutrient status with crop to determine whether the crop needs fertilization. The optimum values are pre-standardized so as to make recommendation after sample analysis.

Diagnosis and Recommendation Integrated System (DRIS)

It is a method of diagnosing nutritional factors limiting crop production, was developed by Beaufils in 1973. The elemental composition of dry biomass of plants consists above 90% of carbon, hydrogen and oxygen, the remains of nutrition composition is made of other essential nutrients to plants, such as: nitrogen, phosphorus, potassium, calcium, magnesium, sulphur, boron, zinc, iron, manganese, nickel, silicon and other elements.

DRIS and critical level approach are used for evaluating nutrient deficiencies. The average index value for the nutrient in the leaf tissue is considered for DRIS. The DRIS norms can provide guidelines for the policy makers in the region regarding recommendations for appropriate fertilizer application. DRIS enables the evaluation of the nutritional balance of a plant, ranking nutrient levels in relative order, from the most deficient to the most excessive.

To be feasible the use of DRIS to assess the nutritional status of plants, the first step is establish the DRIS norms or standard. The DRIS norms consist on average and standard deviation of dual ratio between nutrients (N/P, P/N, N/K, K/N, etc.) obtained from a crop reference population. It is necessary that the crop reference shows high yield.

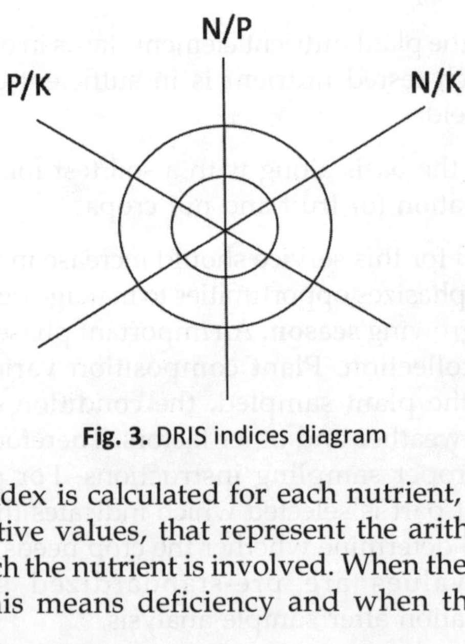

Fig. 3. DRIS indices diagram

The DRIS index is calculated for each nutrient, which may have positive or negative values, that represent the arithmetic average of functions in which the nutrient is involved. When the result is negative (below zero), this means deficiency and when the positive value indicates excess

Index value for N = [(f)*(N/P)+(f)*(N/K)]/2, where f is the function of these nutrients which is calculated by use of coefficient of variation and standard deviation.

Soil Nutrient as an Index of Soil Fertility

Generally, the soil testing laboratories use organic carbon as an index of available N, Olsen's and Bray's method for available P and neutral normal ammonium acetate for K. In semiarid tropics, nitrate nitrogen is also used as an index of available N in soil. Available nutrient status in the soils is generally classified as low, medium and high which are generally followed at the National level and are as follows.

Soil Fertility Ratings

S.No.	Soil Nutrients	Soil Fertility Ratings		
		Low	Medium	High
1	Organic carbon as a measure of available Nitrogen (%)	< 0.5	0.5-0.75	> 0.75
2	Available N as per alkaline permanganate mehod (kg/ha)	< 2.80	280-560	> 560
3	Available P by Olsen's method (kg/ha) in Alkaline soil	< 10	10-24.6	> 24.6
4	Availble K by Neutral N, ammonia acetate method (kg/ha)	< 108	108-280	> 280

Secondary nutrients and critical limits

Element	Soil Test Method	Critical level in soil	Critical level in plant
Sulphur	Hot water, $CaCl_2$ or phosphate	Usual 10 ppm range 8-30	< 0.15 - 0.2%
Calcium	Ammonium acetate	< 25% of CEC or < 1.5 me Ca/100g	< 0.2%
Magnesium	Ammonium acetate	< 4% of CEC or < 1 me Mg/100g	< 0.1-0.2%
Zinc	DTPA	0.6 (0.4-1.2) ppm	< 15-20 ppm
Manganese	DTPA	2 ppm	< 20 ppm (10-30)
Copper	DTPA or Ammonium acetate	0.2 pm	< 4 ppm (3-10)
Iron	DTPA Ammonim acetate	2.5-4.5 ppm	< 50 ppm (25-80)
Boron	Hot water	0.5 ppm	< 20 ppm
Mlybdenum	Ammonium oxalate	0.2 ppm	< 0.1 ppm

□□□

Chapter 4

Soil Organic Matter

Role of Soil Organic Matter and Its Fractionation

It is generally accepted that soil organic matter (SOM) has beneficial effects on soil biological, chemical and physical properties, which in turn influences the productive capacity of soils. It is also accepted that SOM is a major contributor of N, P and S as well as other nutrients to plants. Soil microbial activity is also dependent on SOM as a carbon source for metabolic activity which in turn influences nutrient fluxes and soil structure. The SOM content of soil is generally found to decrease rapidly following the clearing of native vegetation for subsequent cultivation and cropping. Although total organic carbon of the soil declines due to cultivation, of particular concern is the decline in the more labile carbon fractions which are associated with soil nutrient dynamics, as well as having a role in the stabilisation of soil structure. The rate of organic carbon decline varies with management practices, soil type and climatic conditions. Soil aggregation and aggregate stability arc the main factors affecting the susceptibility of soil to raindrop impact and consequently surface sealing and soil erosion. To develop more sustainable cropping systems, management systems which maintain and even improve soil organic matter levels need to be developed. An understanding of organic matter decomposition and its effects on nutrient and carbon balances and soil structure is essential in order to develop sustainable systems.

Soil organic matter (SOM) fractionation

SOM is composed of a number of fractions ranging from very active (labile) to stable (non-labile). According to the ^{14}C dating techniques, the stable organic carbon fraction may have a turnover time as long as several thousand years while the active pool has a turnover time of less than a few decades. Physical fractionation into various components based on size or density has long been used to study the dynamics of SOM.

Soil organic matter (SOM) can be subdivided into non-humified and humified material. Non-humified substances are slightly altered organic materials after decay of tissue from living organisms. Humified substances are decomposition products include complex compounds such as humin, fulvic acid (FA), hymatomelanic acid, humic acid (HA).

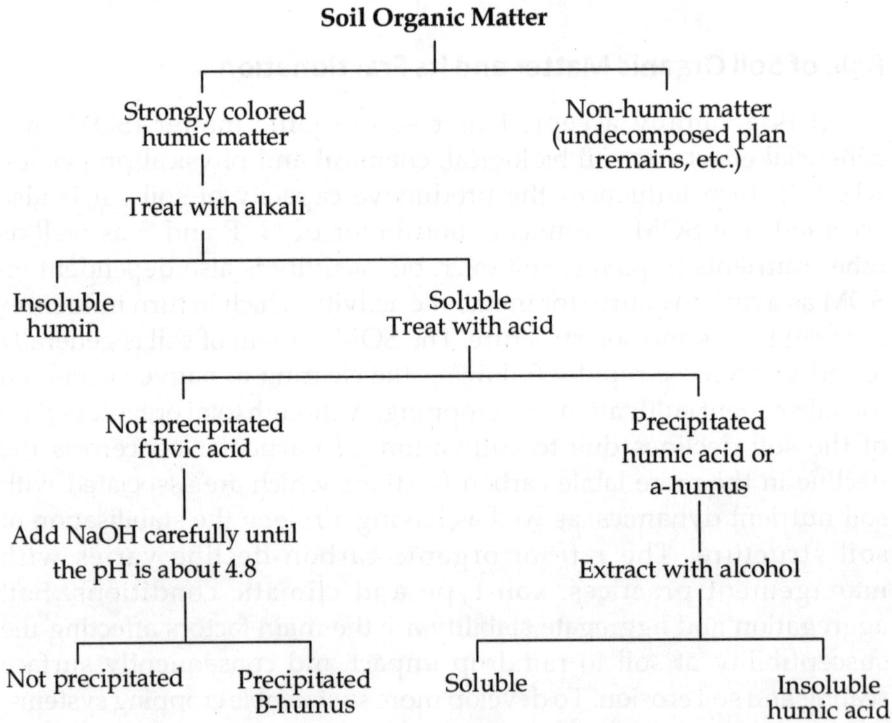

Fig. Fractionation of soil organic matter

The interactions between SOM and soil minerals have been investigated predominantly by various dispersion and separation procedures. The realization that simply shaking soil in water would not ensure complete disintegration of micro-aggregates, led to soil dispersion techniques which utilised various dispersion energies.

Soil Organic Matter Role

- Consists of living organisms, dead plant and animal residues.
- Is the most chemically active portion of the soil.
- Is a reservoir for various essential elements.
- Contributes to CEC.and Buffers soil pH.
- Promotes good soil structure.
- Promotes good air and water relations in plants.

Fate of Organic Material added to Soils

- Undergoes decomposition (only bio-degradable materials) through soil macro and micro fauna and flora.
- Final degradation is made by soil microorganisms.
- The final product is a humus type material which has a C:N ratio around 9:1 - 12:1.
- All complex carbon products are converted into simple compounds

There is a strong relationship between Organic Carbon (OC), Organic Matter (OM) & available N and also there is inter conversion in soil.

Relation of OC to OM and to N

5% of OM is total N

20% total N is available nitrogen

In soil if it has 1.0% of OC then OM is 1.723% i.e., 0.085% total N

$\therefore$ 0.017 % is available nitrogen in soil, then

For 10^6kg soil $= \dfrac{0.017 X 10^6}{100} = 170$ kg per acre

❑❑❑

Chapter 5

Nutrient Cycling in Soils

Fate of Nutrient Elements in Soil

Nutrients are lost through Crop removal, Erosion, Leaching, Volatilization, evaporation, Denitrification and Fixation.

Crop / plant removal

- Plant species have specific requirements of plant nutrients.
- Nutrient removal depends on Growth condition, Crop health, Cultivation and Yield obtained
- Grain crops require more nitrogen than other nutrients. Pulses require more phosphorous while crops such as tomato, banana and pineapple require more potassium compared to other nutrients.

Erosion

Entire top soil is lost through erosion by water or wind; this results in loss of soil phosphorus. Universal soil loss equation can be used here (Soil loss = RKLSCP, tons/ha/year)

Leaching

- Water percolating through a soil profile carry dissolved nutrient elements. Nutrients are easily lost in humid regions and sandy soils.
- Bare soil loses more nutrients than cultivated soils.

Volatilization

Nitrogen is easily lost through volatilization as ammonia, particularly in paddy soils and upland soils in poorly drained areas. This is referred to as ammonia volatilization. This loss is enhanced by high temperature and wind.

Nitrogen Cycle

This is one of the most important naturally occurring events. It is discussed under:

- Nitrogen fixation.
- Conversion of N in soil.
- Mineralization

a) Nitrogen fixation

- Nitrogen is a very inert gas which constitutes 78% of the atmospheric air; it has low chemical reactivity.
- For use by plants it has to be converted to ammonium (NH_4^+) or nitrate (NO_3^-)
- Only a few microorganisms can utilize N gas which is referred to as N fixation.
- Fixation is effected by microbial action, industrial synthesis and high thermal combustion and lightning.

b) Conversion of N in soil

N occurs in soils in the following form :

- Soluble mineral forms, ammonium, nitrate, nitrous oxide (gas)
- Soluble organic compounds, urea, aminoacids.
- Living organisms, plant roots, fungi, bacteria, soil animals.
- Insoluble forms, organic nitrogen, ammonia bonded to clay.

Transformation between the different forms is mediated by soil microorganisms.

c) Mineralization

- Conversion of N in organic residues and soil organic N into soluble forms through mineralization.

- Carbon sources are degraded sources of energy.
- N in excess of microbial need is liberated.

The following sequence of reactions takes place.

i. **Ammonification:** Complex protein compounds are broken down to ammonium compounds by micro-organisms.

ii. **Nitrification:** Ammonium compounds are oxidized to nitrite and to nitrate by two specific types of soil bacteria, *Nitrosomonas* and *Nitrobactor*.

iii. **Denitrification :** The nitrates are reduced to nitrogen gas under poorly aerated conditions through specific micro-organisms. Nitrate form of N is lost through denitrification where nitrogen gas or nitrous oxide is released. This loss occurs mainly in paddy soils and in upland soils which are saturated with water periodically or part of the time.

Fixation

- Takes place by conversion of a nutrient to an unavailable form.
- Phosphorus is converted to unavailable forms both in acidic and alkaline soils, as Al/Fe phosphate or Ca3 (PO4)2 respectively.
- Potassium and ammonium N can be fixed by certain clay minerals.

The Fertility of the Forest

The fertility of the forest, in contrast to other crops in agriculture, is stored not only in the soil, but also in the trees and in the litter of forest detritus on the soil surface. Efforts to utilize forests for additional products (for example, forest debris for fuel) will thus mean an increased removal of fertility. Predicting the magnitude of this loss requires an understanding of the fertility of the forest's various components. Soil in the forest serves as the source and storehouse for elemental nutrients which are needed by trees during growth.

Forest Soil Science: It is the application of all the branches of soil science which deals with the forest soils. How forest maintain the soil fertility and remove the nutrients should be studied to understand the local facts and manage efficiently.

Fertility storage in forest soils

Maintaining soil fertility to trees on the site without losing it through erosion or in water leaching the soil profile is a major problem in forest management. Carbon, although not a nutrient element, represents the soil organic matter that is the basis of most soil fertility storage in forests. Carbon storage is lowest on sites of low productivity, ranging from 10 kilograms per square meter to a depth of one meter to 30 kg per square meter on the most productive sites. This large amount of carbon in the forest soil is a living reality including fungi, bacteria, small soil organisms, and the organic compounds of past generations of forests. This decomposition returns the fertility of past generations to the present tree crop.

Total nitrogen storage in the forest soil, in relation to site productivity, parallels carbon content. On poor sites the average amount stored is 400 grams per square meter and on the highest sites it is up to 1200 grams per square meter to a meter depth of soil. Not all of these large quantities of nitrogen are immediately available to current growth of forest trees because most is bound in the soil's living organisms that decompose soil organic matter.

Phosphorus availability is a problem in forest soils because it may be limited by forming insoluble compounds, or it may be lost through leaching if it is in too soluble form. Furthermore, phosphorus contents may be low even in the best sites. This is probably for different reasons. At the poor sites it may be unavailable, while on the better sites, there may be depletion due to rapid uptake by the trees.

The amounts of both calcium and potassium are highest in the intermediate productivity classes and lowest in the lowest and highest productivity classes. This is probably because of depletion by the tree growth in the more productive sites and a lack in the lower classes. Amounts of magnesium, on the other hand, are greatest in lower productivity classes, and decrease as productivity increases. These elements are stored in the forest soil on the soil's exchange capacity in a form available to the trees. The exchange capacity of forest soils has been found to range from 40 to 300 milli equivalents of exchange capacity per 100 g soil. These values are similar to those in commercial ion exchangers used for water softeners. This reactive capacity of the forest soil, spread over square miles of mountain watersheds, gives us the typical prime soft water quality yield of forested watersheds. As forest trees grow older, a larger proportion of a site's fertility is transferred to storage in stems, branches, and foliage.

Storage in trees and litter

Different quantities of fertility elements are stored in trees and in litter on the forest floor, as well as in the forest's soil. Depletion of soil magnesium could be recharged from soil mineral matter and ultimately recharge would occur in the natural course of forest processes through the litterfall, and decomposition. However, a complete harvesting of trees 160 years old would remove almost one-half of the site's magnesium fertility. Nearly 1,000 grams of nitrogen in a square meter were mainly in soil and leaf litter; only a small proportion was in each tree, mostly in the foliage and branches. By 100 years of age this forest had redistributed the available phosphorus in the soil so that much was stored in leaf litter or in the tree. Phosphorus fertility lost can be easily replenished with fertilization.

Difference Between Forest Soils and Agricultural Soils

Parameters	Forest soils	Agricultural soils
Organic matter	More (> 15%) in surface	Less (<3%)
Porosity	More	Less (<45%)
Soil structure	Undistributed	Distributed
WHC	More (200%)	Less (<60%)
Microbial activity	More	Less to moderate
pH	Mostly acidic	Wide variation (acidic/ neutral/ alkaline
Physical properties	More important for management	Also more important
Infiltration rate	High	Low
Temperature	Cooler	Warmer
Colour	Darker	Lighter to darker
Moisture	Uniform and high	Not uniform
Fertility	More / very high	Less to high
Structure	Granula and crumb	Wide variation
BD	Less (0.6 – 1.1 g/cc	More (1.2 – 1.8 g/cc)
PD	Less (<2.0 g/cc)	More (2.66 g/cc)

Forest Management Implications

Large proportions of some fertility elements in the forest are contained in trees and in leaf litter. This is especially true of phosphorus and nitrogen, and less so of various other mineral elements. Larger amounts of fertility are stored at the tops of the tree trunks, in the branches, the twigs and the foliage; small amounts are in the main trunk. Demands for increased utilization of this forest material will greatly drain fertility from the forest in contrast with past harvests which were confined to the larger trunks. Thus, with intensive timber harvesting, more attention must be paid to replenishing site fertility.

❑❑❑

Chapter 6

Manures and Fertilizers

Manure

Manures are the plant animal waste products which are organic in nature, capable of supplying plant nutrients in available form , bulky in nature having low analytical value and having no definite composition. Formerly, the word in use was "DUNG "in English (at present fertilizer). It is derived from the old German word "TUNG" meaning storage pit covered with manure for protection of soil against cold. From Tung "TUNGEN is derived (to cover) and then dung (fertilizer) as the cover . The corresponding English word "Fertilizer" which is now accepted internationally in view of the old English word DUNG was derived from Latin word "Fertil".

The bulky organic manures are organic in nature and are required to apply in large quantities supply plant nutrients in small quantities, viz, FYM, Composts, Green manures, Sewage sludge. Among these bulky organic manures, FYM, Compost and Green manures are widely used. The effect of bulky organic manures on soils is through the following:

I. Since these manures contain plant nutrients, they have direct effect on plant growth like any other commercial fertilizer. Bulky organic manures contain nutrients in small quantities, therefore large quantities of these are needed. Besides the major nutrients, they also contain traces of micro-nutrients.

II. Bulky organic manures increase organic matter content and hence improve the physical properties of soil. This effect is very important in case of most of our arable land. Such manures increase the humus content of soil and consequently water holding capacity of sandy soils also increased and the drainage of clayey soils is improved .

III. Bulky organic manures provide food for soil microorganisms. This increases the activity of microbes which in turn help to convert unavailable plant nutrients into available form.

Organic manures have been the traditional means of improving soil fertility. The organic matter that is applied thorough organic manures has very complex effect on soil and on plant growth.

Compost

Dry and fresh plant tissues and also animal wastes can be introduced directly into or on the soil. .They can be decomposed at varying speeds, Process known since ageold to convert such wastes into useful agents for soil improvement i.e., compost. The word compost is derived from the Latin word "COMPONERE" to mean put together. The compost is a product of decomposition of plant and animal wastes with various additives .

Composting : Composting is a process of converting organic matter in to manure in a short time by accelerating fermentation process under controlled conditions is called composting. The organic material that is usually handled for composting are waste vegetative material such as spoiled straw, peanut hulls, saw dust, dried leaves etc., stubbles, chilli plants, cotton stalks, tobacco stems, weeds, municipal waste depending on their availability.

Green manuring: Green manuring can be defined as a crop for the manuring purpose by incorporating it into soil while green, or soon after maturity with a view to improve the soil and benefit subsequent crops (*In-situ*). It is also applied from outside source (*Ex-situ*).

Fertilizers

Fertilizers are those mined or manufactured chemicals which are used to supply nutrients to the plants. Fertilizers are the compounds that are added to plants to promote growth. There are two types of fertilizers-organic and inorganic. Organic fertilizers are carbon based and are composed of organic matter like leaves, cow dung and parts of plants. And inorganic fertilizers are those synthesized chemically viz., Urea, DAP, MPO, SSP, I7: I7: I7, I9: I9: I9, I2:32:I6, I0:26:26, 20:20:20.

A fertilizer can be defined as a mined or manufactured material containing one or more essential plant nutrients in potentially available forms in commercially valuable amounts.

Classification of Fertilizers:

I. Based on nutrient content

Single nutrient and more than one nutrient fertilizer

II. Based on physical mixing Mixed Fertilizers: and CGF - Customized granulated fertilizers

III. Based on complexity

- Simple / straight fertilizers : one nutrient
- Complex fertilizers: >one nutrient and have mixtures of chemicals in reaction

Straight fertilizers: It is required to know percent nutrient content in any fertilizer. One should know how it is worked out.

Fertilizer	Working out content
Ammonium Nitrate (NH_4NO_3) N-14X1=I4 H- 1X4=4 N- 14X1=14 O- 16X3=48	Fertilizer 80 g contain - 28 g N Therefore 100 g contain 35%N
Ammonium sulphate $(NH_4)_2SO_4$ (I4+4)*2+32+4*I6 Urea {CO $(NH_2)_2$} 12+16+(14+2)*2 Muriate of (Potash KCl) 39+35.5	Fertilizer 132 g contain - 28 g N Therefore 100 g contain 21%N Fertilizer 60 g contain - 28 g N Therefore 100 g contain 46% N Fertilizer 74.5 g contain - 39 g K Therefore 100 g contain 60% K_2O Here k to K_2O factor is 1.2
Single Super Phosogate (Ca_2HPO_4) 40*2+1+31+64	Fertilizer 176 g contain - 31 g P Therefore 100 g contain 16% P_2O_5

Complex fertilizers

DAP - It is one of the complex fertilizer used to supply phosphorous to crops. It contains I8% N and 46% P_2O_5.

The other complex fertilizers contain I7:I7:I7, I9: I9: I9, I0:26:26, 12:32:16, and many more and indicate NPK (% N, P_2O_5, K_2O).

The NPK fertilizers also produced by Physical Mixing. Basic principle is to melt crop recommendation using strait fertilizers, they all mixed to get required ratio to meet crop need. According to soil status mixing the fertilizers is also done.

Example with RDF 40:20:20 to paddy crop is taken here. First what are the fertilizers available in the local area should be taken into account. Suppose Urea, SSP and MOP are available, then the quantities required to supply RDF will be :

Urea (with 46% N) will be 86 kg

SSP (with16% P_2O_5) will be 125 kg

MOP (with60% K_2O) will be 33 kgm

Classification of Fertilizers Based on Type of Nutrients

1. Nitrogenous Fertilizers
2. Phosphatic Fertilizers
3. Potash Fertilizers
4. Micro-nutrient Fertilizers.

Based on application

I. Soil application:

- Dry Fertilizers: granulated powder, prilled frits
- Liquid Fertilizers: nutrient content mixed with water
- Gaseous Fertilizers: normally NH_3 is used
- Drip Fertilizers: along with H_2O it is applied for irrigation-fertigation.

II. Foliar application

- Nutrient supplied through foliage. Here water soluble fertilizers are used.

Based on type

I. **Organic Fertilizers:** obtained from organic sources i.e., plant and animal residues

II. **Inorganic Fertilizers:** chemical sources and contain large amount of nutrient content

	Organic	Inorganic
Content	Less	High
Quantity	Bulky	Small
Release of nutrient	Slow	Quick
Cost	Highly expensive (per unit nutrient)3000-4000 Rs./ton (1 ton contains N- 2% P- 0.5% and K- I%)	Less costly to organic based on nutrient content It contain high NPK content (single or one)

Nitrogenous Fertilizers

First nitrogenous commercial fertilizer is Guano it is the dropping of birds. Later on $NaNO_3$ is mind and used for N fertilizer at chilli in South America.

Eg: Urea, Ammonium sulphate, Ammonium Nitrate.

Nitrogenous Fertilizers are produced by using NH_3.

NH_4NO_3 (33%), Nitro phosphate (I2-20%), $(NH_4)_2SO_4$ (2I%), Urea (46%), Liquid Ammonia (20%), Aqua Ammonia

Urea is produced by Haber and Busch process.

Here $NH_3 + CO_2$ – – – – – – – Urea (Prilling)

Electric Are ↓

Under cold conditions granulation is obtained

Advantages of Urea:

- It is convenient in all soils for application and for all crops
- Used in small quantity because of its high 'N' content
- Foliar application is possible
- Easy for broadcasting

- It can be used for basal and top dressing.(but P and K only for basal dressing)
- Urea is less acidic compared to other ammonical fertilizers.

Urea about I2.6 mt of nitrogenous fertilizers are produced in India. The industries are in Bombay, Mangalore, and Goa etc….

Slow release N fertilizers

In Urea fertilizer release takes place hence there will be nitrogen losses by 3 ways.

i. Volatilization (NH_4)

ii. Leaching (NO_3)

iii. Denitrification (N_2)

To reduces these losses treat coal with urea – "Urea Tar"

N serve is used, the chemical which reduce the oxidation of Urea

Phosphatic Fertilizers

P fertilizers mainly produced from Rock Phosphate [$Ca_3(PO_4)_2$] called as Apatite. P is required for energy molecule (ATP, ADP) in human and plants and for microbe's.

$Ca_3(PO_4)_2 + H_2SO_4 \rightarrow$ Single Super Phosphate (SSP) contain 16 % P_2O_5

$Ca_3(PO_4)_2 + H_3PO_4 \rightarrow$ Ammonium Phosphate (has 20-54% P_2O_5)

$Ca_3(PO_4)_2 + H_3PO_4 \rightarrow$ Tripe Super Phosphate (TSP) (has 42- 50% P_2O_5)

$Ca_3(PO_4)_2 + HNO_3 \rightarrow$ Nitro Phosphate contain 11- 35% P_2O_5

- This $Ca_3(PO_4)_2$ is of sedimentary origin
- Naturally available and it is excavated from the place where it is found (in hillocks)
- This rock is phosphate is more in Ethiopia, China and Russia
- Nowadays all these fertilizers are produced through wet process, i.e., by use H_2SO_4, H_3PO_4, HNO_3
- In 1847 its production was around 500 tons
- In 2013 its production has increased to 67000 billion tons
- In India the major R/P source is in Rajasthan and MP

Rajasthan produces 80-85% of rock phosphate required by India. The demand in India is 10 m.t but there is a production of 1.5 m.t only. In Rajasthan there is a reserve of above 80 m.t

In Madhya Pradesh there is a reserve of rock phosphate above 51 m.t. So, to meet the demand it is required to import the rock phosphate from Egypt, Bangladesh, Morocco, US, China, Iraq

- In older times i.e., before 1980 rock phosphate rate was very low (i.e. Rs. 1000/- per ton). It has increased to Rs.5400/t, in 2009 (i.e. to 90$)
- Nw the existing rate is Rs. 7400/t, in 2014. (185$)

Potassic Fertilizers

MOP (Muriate of Potash) - KCL - contain 60% K_2O or 50% K

SOP (Sulphate of Potash) - K_2SO_4 - contain 50% K_2O

K to K_2O conversion = > Kx1.2 and K_2O to K conversion => K_20 /1.2

Potassium Nitrate - KNO_3 contain both K and N (almost 39 and 14 %)

The potash fertilizer is not produced in India. The K reserves are more in Canada i.e. in Saskatchwan area. It is imported.

- Tigray's of Dallol in Europe- about 150 mt of K reserves are present
- There are more reserves in China, Russia, Brazil even north America, Belarus countries
- In 2011 there was a production of 37 mt of potash in the world
- The price of MOP fertilizer in 2008 - is 200$ per ton
- The price of MOP fertilizer in 2013 - is 393$ per ton

Customized Granulated Fertilizers (CGF)

In the market complex fertilizer: I7: I7: I7 is sold at Rs.2500 per q. But same thing can be prepared by the method CGF with low cost.

To produce one ton of fertilizer 17:17:17 by using DAP, Urea and MOP, the quantities will be :

To get 170 kg phosphorous, the DAP (with18%N and 46% P_2O_5) will be 369.5 kg

Here the nitrogen from DAP has to be deducted from requirement.

To get remaining 104 kg Nitrogen,the Urea (with 46% N) will be 226kg

To get 170 kg potash, the MOP (with60% K_2O) will be 283 kg

By adding all (226+369.5+283) we get 878 kg. To make it a ton, remaining will be the proposed filler i.e. I22kg. Therefore I000kg will be prodused with the help of filler. It is required to add Bantonite clay to fill the remaining quantity as filler.

Let us work out the cost

During 2015, the cost of 17:17:17 complex fertilizer in the market was Rs.25000 per ton. By using DAP, Urea and MOP, the same grade can be produced by CGF unit. The cost will be :

Urea (Rate Rs.900 / q) for 226 kg it will be Rs.2034

DAP (Rate Rs.2000 / q) for 369 kg it will be Rs.7400

MOP (Rate Rs.600 / q) for 283 kg it will be Rs.1698

Filler (Rate Rs.2000 / ton) for 122 kg it will be Rs.244

i.e the cost of Urea = Rs. 2034 for 226kg

DAP = Rs. 7400 for 370kg

MOP = Rs. I698 for 283kg

Filler = Rs. 244 for I22kg

The total cost of raw materials is Rs. 11376/-

Process cost (labour, electricity, other materials) it is 50% of actual cost i.e. Rs.5688 per ton. And the total cost will be Rs.17064 per ton. Nearly Rs.8000 per ton will be available as profit and marketing margin. With 10.0 tph capacity, production with 6 hours a shift, in 3 shifts a day, 25 days in a month, 8 months in a year, if done, the total quantity of fertilizer produced will be 36000 tons per annum. Now you can imagine the turn over and margin money. Like 17:17:17 fertilizer, other grades can also be produced at lower cost.

Flow of materials in CGF manufacturing plant

Materials will move from :

Hopper → paddle mixture → Bucket Elevator (BE) → 3 Drums→ B.E. → Screen (> 4mm < I mm will be recycled)

In first drum moisture is added, and then in another drum heat is supplied. In third drum cooling of granules will be made. Drums will be moving in rotary action with hammering on surface to loosen the adhered material inside.

The CGF Plants with different capacities: 2.5, 5.0 and 10.0 tons per hour (approximate cost of machinery in 2015 was Rs.2.5 to 5.0 Crores). According to the capacity, it is required to feed the fertilizer quantities as raw materials.

Suppose plant requires 10 tph, we have to feed for 17:17:17 fertilizer

2260kg Urea/ha = 9.4kg/15sec

3700kg DAP/ha = 15.29/15sec

2830kg MOP/ha = 11.7kg/15sec

1220kg Filler/ha = 5.08 kg/15sec

Fillers are the non-nutrient materials added to the fertilizers to make the bulk quantity.

Advantages of CGFs

- It requires less labor for application
- It has balanced nutrients. N, P, K
- Micro-nutrients can be added
- Easy for application because CGF (nutrients) are in granule from

Disadvantages

- Not able to apply same for all the crops
- Individual nutrient application is not possible

Care must be taken while mixing fertilizers – as physical

a. Ammonical fertilizer which contain (NH_4): should not mixed with lime, it forms alkalinity, volatilization loss will be there.

b. Urea should not be mixed with SSP, Urea attracts the moisture; if it is mixed with SSP it forms cake.

c. SSP should not be mixed with lime. Otherwise P reacts with Ca and forms rock phospahte.

d. $NaNO_3/KNO_3$ should not be mixed with SSP because it also forms cake.

In order to avoid this, we can add fertilizer conditioners.

Organic materials (powder form) added to fertilizers during the process of production to reduce caking (75kg/ton).

Eg: paddy hull, groundnut hull, tobacco leaves/ stem, neem cake etc.

Nutrient ratio: It is the nutrient content in 100 units of the fertilizer, i.e., 10:26:26 NPK in 100.

Fertilizer Grade: it is the minimum guarantee of nutrient content in percentage (%).

Fertilizer Grade

There are 5 types of fertilizer grade

Grade I	N&P α 6%N and 12% P_2O_5, 0% K_2O	Grade II	10%N, 6% P_2O_5, 6% K_2O
Grade III	6.6%N, 6% P_2O_5, 6% K_2O	Grade IV	N & P are equal (9.0% N & 8.8% P_2O_5)
Grade V	N, P & K are equal (8%)		

These grades are used for different crops

Grade I	Used for grain crops	Grade II	Potato, vegetables, banana, papaya
Grade III	Coconut, arecanut	Grade IV	Sugarcane, potato, and grain crops and groundnut
Grade V	Potato, oil seeds, cereals		

Fertilizer ratio

Based on soil test, suppose if it is required to apply 100:50:75 kg NPK/ ha, then the fertilizer ratio is 4:2:3.

Secondary Nutrient Fertilizers

- Ca, Mg, S – are secondary nutrients

  Lime- $CaCO_3$

  Dolomite – $CaCO_3$, $MgCO_3$

  Gypsum – $CaSO_4$, $2H_2O$

Sulphur - Bentonite sulphur

Megnesium Sulphate - $MgSO_4$

Sulphate forms of metallic micronutrients also supply the sulphur.

In market in the name of 'Set-rite' we get the product which may have gypsum, dolomite or lime

- Wherever there is reddening in leaf in case of Mg deficiency use $MgSO_4$ (1% or 2%) which acts as spray fertilizer and also soil application fertilizer
- Even supply of S is also done through $MgSO_4$

Micro Nutrient Fertilizers: B, Mo, Cu, Fe, Mn, Zn. The B micronutrient is supplied from Borax, 3 to 4 kg/acre which contains 10.5% B which is used for soil application.

- Borosol or Solubor contains 20% Boron which can be sprayed at 0.2%
- Mo- $NaMoO_4$ (Sodium Molybdate) or NH_4MoO_4 (Ammonium molybdate). it is used for spraying or seed treatment at 0.01% because Mo is very-very toxic while major nutrients / secondary nutrient, are sprayed at 0.5 - 2%
- The other micronutrients Cu Fe Mn Zn are available in sulphate forms around 21%. In market they are available in EDTA forms also wherever the soil condition is not good, then this organic molecule EDTA is used as complexing agent.
- Boronated SSP : here 0.18% Boron is added to SSP (Single Super Phosphate)
- Zincated SSP : 2.5% Zn is added to SSP

Liquid Fertilizer

- These are solutions prepared from soluble fertilizers
- These are used as starter doses
- Can meet the recommended fertilizer up to 10%

Water Soluble Fertilizers

- 100% soluble fertilizer

- Soil application fertilizer- here soil, sand and other materials are also present and not 100% soluble, there filtering of solution is taken up.

In water soluble fertilizer there are 2 grades

i. Spraying grade- spraying purpose [DAP of 2% prepared and sprayed]

ii. Drip grade – for drip [soluble fertilizer are used at 0.5 - 2.0%]

Fertilizer Application: Major nutrients N,P,K

N needs to be applied as Basal and Top dressing

P needs to be applied as Basal

K Normally all dose as Basal, under certain conditions like sandy soil, it is used as top dressing besides basal

N needs to be applied in applied in split doses for avoiding de-nitrification, leaching and volatilization.

Nutrient Removal

It is nothing but nutrient uptake from soil per unit of crop produce

One quintal paddy requires 2 kg N which can taken as nutrient removal (kg / q of grain). One ton of paddy straw requires 20 kg each of N and K (kg / t of straw).

Nutrient Budgeting

It is calculated from the nutrient Arrival and Expenditure. It will be Deficit when arrival is less then expenditure.

There are nutrient budgeting components such as

- Nutrient available in soil and applied quality
- Nutrient required by crop
- Nutrient uptake by crop
- Nutrient balance in soil
- Under dry conditions fertilizer application should be made slightly deeper (Especially Phosphorous) because of less moisture and to favor root growth.

- Along with the seed, fertilizer is normally not mixed because there will be salt injury to the germinating seed.
- P fertilizer should be applied in bands, placed 5-10cm deep, because there will be more contact of P-fertilizer with soil.

Potassium

It should be applied as basal. It is susceptible for leaching and also soil particles there will be fixation of K

There will be non- exchangeable K, exchangeable K soluble K, in equilibrium form in soils.

Non-exchangeable → exchangeable → soluble K

Synergistic Effect

It means positive interaction. The nutrients NP, NK have this effect. One uptake will the uptake of the other.

The application of one fertilizer will influence the uptake of other that helps in internal growth.

Antagonistic Effect

It means –ve interaction

The application of one nutrient will hinder the uptake of the other.

Examples:

I. Excess of P application cause deficiency of Zn, Fe, Cu,

$Ca(HPO_4)_2 + Fe(OH)_3 \rightarrow 2Fe(H_2PO_4)_2 + Ca(OH)_2$

Mono Calcium Phosphate (MCP)

$Ca(HPO_4)_2 + A1(OH)_3 \rightarrow A1(OH)_2H_2PO_4 + Ca(OH)_2$

II. Heavy N application cause deficiency of Cu

III. Over liming induces B and K deficiency

$CaCO_3$+B Calcium Borate (insoluble)

Ca $K^+ \rightarrow$ leaching losses may appear.

IV. Excess N application induces Mn deficiency

V. Excess of Mn, Zn, Cu causes Fe deficiency

VI. Excess of Fe, Zn, Cu, causes Mn deficiency

VII. Excess of lime converts available P into TCP(Tri Calcium Phosphate)

$Ca(HPO_4)_2 \rightarrow Ca_3(PO_4)_2 + 2CO_2 + 3H_2O$

MCP

VIII. In I:I clay P is going to be fixed because of OH ions and P^H dependent charges.(I:I-1Si:1A layers)

Biuret: it should be < I % in any fertilizer normally in Urea. It is poisonous to plants

Urea + Urea → Biuret with an exclusion of H_2O molecule

Liming affects volatilization of NH_4

$NH_4C1 + lime \rightarrow CaCl_2 + NH_3\uparrow + CO_2\uparrow + H_2O$

(soluble)

$NH_4(SO_4)_2 + CaCO_3 \rightarrow CaSO_4 + NH_3\uparrow + CO_2\uparrow + H_2O$

(slowly soluble)

$NH_4NO_3 + CaCO_3 \rightarrow Ca(NO_3)_2 + NH_3\uparrow + CO_2\uparrow + H_2O$

(soluble)

In calcareous soils, lime is not used because of the volatilization of NH_4 is dominant in these soils. Under alkaline pH, in calcarious soils, high Ca and Mg are present. Some other causes are:

i. Low CEC (I:I clay) cause N loss
ii. High soil temperature and evaporation also cause N loss
iii. Less moisture also cause N loss
iv. Application of more quantity of ammonical fertilizer also cause N loss
v. Nitrogen undergoes mineralization and immobilization in soils

Mineralization	**Immobilization**
Organic N is converted in to inorganic N (NH4+and NO3-)	Inorganic N is converted to organic N but it is temporary shock to plants. More of C – lignin, cellulose, it applied to soil
Available to plants	Unavailable to plants for sometimes

K- Fixation

It goes permanent to non-exchangeable form.in soils. Inbetween layers of silicate minerals the potassium will be fixed. It is common in Illite clay, Vermiculite clay & Chlorite clay which are specific K fixers. They highly fix K. Montemorollite clay adbsorb K on exchange sites.

How K imparts disease resistant in plants?

K improves lignifications process which hinders the entry of pathogens.

(Lignin means hard cell wall, cell wall becomes thicker).

C:N Ratio

In any Organic Matter, the carbon is I0 to 20 and N is 1. The Wide C:N ratio 20:1 (i.e. they are hard materials with more C) makes decomposition slow.

Narrow C:N ratio I0:1 (soft materials with less C and more N). The C:N ratio of I0:1 is favorable for decomposition of organic matter **otherwise** make decomposition slow.

Need of Fertilizers Application in Forest Plants

Normally fertilizers are not applied for forest plants as in agricultural crops because fertilizers are obnoxious, hazardous, not eco-friendly and they are not economical In forestry fertilizers are not advised except for nursery (for improvement of growth).

Then how to fulfill the requirements?

Supply of nutrients in forest plants: if we supply organic fertilizers there will be slow release so it is better to use bio-fertilizers (are the microbes used as a nutrient suppliers) along with organics.

Eg: N—fixation bacteria
- Free living
- Associative

} atmospheric N is fixed from source

VAM- Vesicular Arbasicular Mycorrhizae

- VAM is an endomycorrhiza
- In ectomycorrhiza on root surfaces there are nodes the Hyphae enter into cortex of the root, it will supply nutrients to plants.

The association of such plants are: Pinaceae family plants, Fugacea, Saliaceae, Eucalyptus, Casuarina, Tilia. Here Pinaceae and Fugaceae have only Ectomycorrhiza and others have both Ecto and Endomycorrhiza.

Ectomycorrhiza belongs to fungi- Basidiomycetes

- It is nothing but a fungi
- It brings nutrients from available source
- There is a record of applying 1000kg of VAM in California for citrus
- It can draw nutrients from faraway places about> 1000km
- VAM belongs to Zygomycetes
- VAM has association with Angiosperms, Gymnosperms, Pteridopgytes, Bryopgyter, Eucalyptus, Casuarina (Pinaceae and Firidaceae)
- This VAM is susceptible for waterlogged condition

VAM is more suitable in forest because:

- Forest plants are slow growing
- Planted in infertile lands
- Has high weathering capacity
- It can supply nutrients to plants even in infertile soil condition. It is going to supply Zn and P.
- It gives strength to plants
- Avoid transpiration shock by supplying water and nutrients.

Azolla, BGA, Azotobacts, Azospirallum, Rhizobium. These are not much useful as that of VAM in forest but are useful in agricultural crops.

Method of Application of Fertilizers

1. Broadcasting
2. Row Placement
3. Spraying
4. Fertigation

I. Broadcast Method

Broadcast application is a means of applying large quantities of nutrients that cannot be conveniently added @ planting. The nutrients are often mixed through the [plow] layer, which helps to provide nutrients later in the growing season. They are more likely to be in a moist zone for a greater part of the year because elements such as phosphorous and potassium move to the roots by diffusion through water films.

Advantages

- Labor saving during planting
- Money and time is saved
- Easy for application

Disadvantages

- Wastage of fertilizers
- There will be evaporation loses of materials
- Along with H_2O leaching, run off may take place under uneven distribution
- It is going to effect the environment because evaporated gases will accumulate in the atmosphere.

II. Row Placement

Suitable for supplying basal dose. Placement of fertilizer is done adjacent to the seed. P fertilizer is applied in bands.

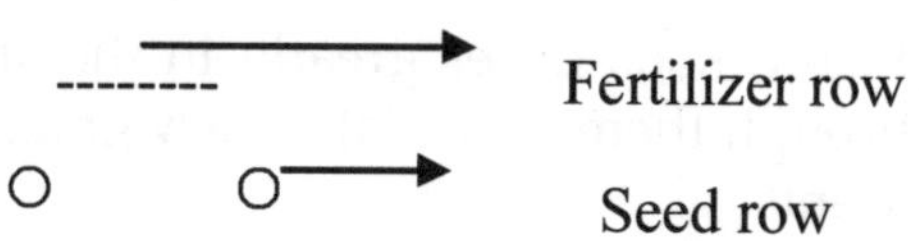

III. Spraying

Fertilizers are given in solution form. This is used for frequent supply of nutrients. These nutrients are supplied whenever there is deficiency. By spraying it is possible to apply 10% of total nutrient.

IV. Fertigation

This is the application of fertilizer in irrigation water in either open or closed systems. The open system includes lined and unlined open ditches and gated pipes that are used for furrow and flood irrigation methods. Sprinkler, spitter, trickle, drop and dual wall turning systems are the main types of open systems.

- Nitrogen phosphorous, potassium and sulphur are the principal nutrients applied by fertigation.

Advantages

- Even distribution of fertilizers
- Labor is saved
- Higher yield
- Midseason deficiencies in crops can be corrected by fertigation

Things required

- Drip system
- Fertilizer tank
- Ventury system
- The maximum stock solution concentration of nutrient in fertigation should be 10%

Time of Application of Fertilizer

The time at which to apply a fertilizer depends on the soil, climate, nutrients and crop.

With respect to the soil factor, soils differ greatly in the speed with which water will move through them. They also **vary** greatly in their capacity to fix plant nutrients.

1. Basal application → P, K, Zn, B, Fe
2. Top dressing → N and K

Balanced fertilization

Balanced fertilization does not mean a certain definite proportion of Nitrogen, Phosphorus and Potash (or other nutrients) to be added in the form of fertilizers and organic manures after taking into account the availability of nutrients already present in the soil, crop requirement and other factors. Generally referred 4:2:1 NPK use as a desired ratio does not substitute the need and importance of actually working out the nutrient deficiency in the soil and addition of required nutrient through fertilizers and manures to meet the crop need. This ratio actually relates to the general fertilizer use recommendation started for the two major cereal crops, i.e. rice and wheat as 120 : 80 : 40 kg NPK per hectare (4 : 2 : 1). Till the beginning of green revolution era, farmers have been traditionally using only nitrogen as the first nutrient being almost universally deficient in Indian soils which also showed conspicuous response to its application. Thus, an emphasis on a particular ratio (4 : 2 : 1) has helped in increasing the use of P&K which was necessary to be used over the application of N when high yielding varieties were introduced. Balanced fertilization should also take into account the crop removal of nutrient, crop species to be sown, farmers' investment ability, soil moisture regimes, weed control, soil salinity, alkalinity, physical environment, microbial status of the soils which determine the status of available nutrients in soil and cropping sequence etc.

The balanced fertilizer should aim at:

I. Increasing crop yield and quality.

II. Increasing farm income

III. Correction of inherent soil nutrient deficiency.

IV. Maintaining or improving soil fertility

V. Avoiding damage to the environment

VI. Restoring - fertility and productivity of the land that may have been degraded by exploitive activities in the past.

Soil Amendment

Amendments are those chemicals which are used for modification of the soil property particularly, pH and physical conditions are required.

$$\text{Soil amendements}\begin{cases}\text{Acid soils} - CaCO_3/CaCO_3.\ MgCO_3\text{- Dolomite}\\ \text{Alkali Soils} - CaSO_4.\ 2H_2O\text{- Gypsum}\end{cases}$$

In Acid soils

$$CaCO_3 \xrightarrow{\text{Sulphur}} Ca(clay) + H_2O + CO_2 + H_2SO_4$$

Sodic soils → Alkali soils → these are rich in sodium. For this soil gypsum is applied and good water is provided for drainage through which Na is removed.

$$\left.\begin{matrix}Mg\\ H\ \ Clay\\ Ca\end{matrix}\right] Na + CaSO_4.2H_2O \rightarrow \left.\begin{matrix}Ca\\ Clay\\ H\end{matrix}\right] \quad Mg + Na_2SO_4 \downarrow$$

(Soluble in H_2O)

Fertilizer Use Efficiency (FUE)

FUE is defined as the efficiency with which the crop yields are maximum with additional use of fertilizer i.e., q/ unit of fertilizer multiplied by 100 (in terms of %).

Fertilizer use efficiency depends on fertilizer and the nutrient in it.

$$\left.\begin{matrix}N \xrightarrow{\text{FUE}} 40\text{-}50\% \\ P \longrightarrow 30\% \\ K \longrightarrow 60\%\end{matrix}\right\} \text{rest will be fixed in soil or lost}$$

Chapter 7

Integrated Nutrient Management

It is essential to adopt Integrated Nutrient Management - INM, to get higher sustained yield by maintaining soil health. It is also called as IPNS- Integrated Plant Nutrient System, which means supply of plant nutrient from various sources (soil reserves and organic manures). It Concerns the maintenance of soil fertility to an optimum level of management for sustaining the crop produce

Soil Reserve

It is the reserve form of soil fertility / nutrient status of soil available on depletion of soluble form. If we apply N, volatilization, leaching and de-nitrification will be there even though some uptake by plants can be seen but soil reserve cannot increase suddenly. Hence, N buildup may not be seen though chemical fertilization. P reserve can be increased by applying P nutrients in excess. Similarly K reserve also can be increased provided there is no leaching.

Organic Manures

The manures from plant and animal origin are referred to be organic and also they contain carbon.

$CO(NH_2)_2$ that is the Urea, though it contains carbon it can not be organic. It is synthetic fertilizer.

Organic manures examples: Poultry manure, FYM, Vermi compost, Fish meal, blood meal, Neem cake, Night Soil are the organic manures and their nutrient contents are given below :

Organic manures	N(%)	P_2O_5(%)	K_2O(%)
FYM	0.5	0.3	0.5
Compost	0.5	0.15	0.5
Night soil	0.5	4.0	2.0
Sewage sludge	6.0	2.0	1.0
Poultry manure	3.0	2.0	1.0
Fish manure	4.0	5.0	1.0
Neem cake	5.0	1.0	1.0
Groundnut cake	7.0	1.5	1.4
Blood meal	10.0	1.0	1.0
Bone meal	2.0	20.0	1.0

→ For only ½ kg of N we need to apply 100 kg of manure (FYM / Compost), while nearly one kg of Urea is sufficient to supply nearly ½ kg of N.

Significance of Organic Manure

- Bulk density of organic manure is low. It releases the nutrients very slow ans it is most favorable for forest plants.
- Quality of produce is improved
- Organic manures are eco-friendly in nature
- Saves the foreign exchange and National economy increases
- Improves soil structure and there five porosity increases and WHC increases.

In INM one more component is green manure

Ex: Serbania used, Dahincha, Glyricidia, Pongamia, Cowpea, Subabul, etc., green leaves are:

- The production of green foliage will be 2-21 t/ha
- Green manure is a rich source of N (2%)
- It is succulent and easily decomposable
- During composting I is used as one of the component.

Bio-fertilizers are micro-organisms which bring about nutrient enrichment of soil by enhancing the availability of nutrients to crops.

The micro-organisms which act as bio-fertilizers are bacteria, cyanobacteria (blue green algae) and mycorrhizal fungi. Bacteria and cynobacteria have the property of nitrogen fixation while mycorrhizal fungi preferentially withdraw min-erals from organic matter for the plant with which they are associated.

Nitrogen fixation is the process of conversion of molecular nitrogen into nitrogen compounds. Insoluble forms of soil phosphorus are converted into soluble forms by certain micro-organisms. This makes the phosphorus available to the plants. Phosphate is also solubilised by some bacteria and by some fungi that form association with plant roots.

Free Living Nitrogen Fixing Bacteria

They live freely in the soil and perform nitrogen fixation. Some of them are saprotrophic, living on organic remains, e.g., Azotobacter, Bacillus polymyxa, Clostridium, Beijerinckia. They are further distinguished into aerobic and anaerobic forms.

The property of nitrogen fixation is also found in photoautotrophic bacteria, e.g., Rhodopseudomonas, Rhodospirillum, Chromatium. Inoculation of soil with these bacteria helps in increasing yield and saving of nitrogen fertilizers. For example, Azotobacter occurring in fields of Cotton, Maize, Jowar and Rice, not only in-creases yield but also saves nitrogen fertilizer to the tune of 10-25 kg/ha. Its inoculation is available under the trade name of azotobactrin.

Free Living Nitrogen Fixing Cyanobacteria

A number of free living cyanobacteria or blue-green algae have the property of nitrogen fixation, e.g., Anabaena, Nostoc, Aulosira, Totypothrix, Cylindrospermum, Stigonema. Cyanobacteria are photosynthetic. Therefore, they add organic matter as well as extra nitrogen to the soil.

Aulosira fertilissima is consid-ered to be the most active nitrogen fixer of Rice fields in India (Aiyer et al, 1972). Cylindrospermum licheniforme grows in Sugarcane and Maize fields. Cyanobacteria are an extremely low cost biofertilisers. In Tamil Nadu, the technique of cyanobacteria inoculation to rice fields is being followed. Phosphate, Molybdenum and Potassium are supplied additionally.

Partial Association of Nitrogen Fixing Bacteria

Certain nitrogen fixing bacteria like Azospirillum live around the roots of higher plants without developing any intimate relationship. It is often called rhizosphere association. The bacteria obtain some plant exudate and use the same as part of their food requirement. The bacteria fix nitrogen and exude a part of the fixed nitrogen for use by the plant.

Symbiotic Nitrogen Fixing Bacteria

They form a mutually beneficial association with the plants. The bacteria obtain food and shelter from plants. In return, they give a part of their fixed nitrogen to the plants. The most important of the symbiotic nitrogen fixing bacteria is Rhizobium. It forms nodules on the roots of legume plants. There are about a dozen species of Rhizobium which form association with different legume roots, e.g., *R. leguminosarum, R. lupini, R. trifolii, R. meliloti, R. phaseoli.*

These bacteria, also called rhizobia, live freely in the soil but cannot fix nitrogen except for a strain of Cowpea Rhizobium. They develop the ability to fix nitrogen only when they are present inside the root nodules. In the nodule cells, bacteria (bacteroids) lie in groups surrounded by membrane of the host which is lined by a pink-red pigment called leghaemoglobin. Presently cultures of Rhizobium specific for different crops are raised in the laboratory.

Frankia, a nitrogen fixing mycelial bacterium (actinomycete), is associated symbiotically with the root nodules of several nonlegume plants like Casuarina, Alnus (Alder) Myrica, Rubus etc. Leaves of a few plants (e.g., Ardisia) develop special internal cavities for pro-viding space to symbiotic nitrogen fixing bacteria, Xanthomonas and Mycobacterium. Such leaves are a constant source of nitrogen fertilizer to the soil.

Symbiotic Nitrogen Fixing Cyanobacteria

Nitrogen fixing cyanobacteria (blue- green algae) form symbiotic association with several plants, e.g., cycad roots, lichens, liverworts, Azolla (fern). Out of these, Azolla-Anabaena association is of great importance to agriculture.

Azolla pinnata is a small free floating fresh water fern which multiplies rapidly, doubling every 5-7 days. The fern can coexist with rice plants because it does not interfere with their growth. In some South-East Asian countries, especially China, the rice fields are regularly provided with Azolla.

Anabaena azollae resides in the leaf cavities of the fern. It fixes nitrogen. A part of the fixed nitrogen is excreted in the cavities and becomes available to the fern. The decaying fern plants release the same for utilization of the rice plants. When field is dried at the time of harvesting, the fern functions as the green manure, decomposing and enriching the field for the next crop.

Microphos Biofertilizers

They release phosphate from bound and insoluble states, e.g., Bacillus polymyxa, Pseudomonas striata, Aspergillus species.

Mycorrhiza (pl-Mycorrhizae Frank, 1885)

It is a mutually beneficial or symbiotic association of a fungus with the root of a higher plant. The most common fungal partners of mycorrhiza are Glomus species. Mycorrhizal roots show a sparse or dense wooly growth of fungal hyphae on their surface. Root cap and root hairs are absent.

The shape is irregular, tuberous, nodulated or coralloid. The fungus remains restricted to the cortex of the root. The vascular strand and growing point are not affected. Mycorrhiza often remains in the upper layers of the soil where organic matter is abundant. Depending upon the residence of the fungus, mycorrhizae are of two types – ectomycorrhiza and endomycorrhiza.

(a) Ectomycorrhiza (= Ectotrophic Mycorrhiza)

The fungus forms a mantle on the surface of the root. Internally, it lies in the intercellular spaces of the cortex. The root cells secrete sugars and other food ingredients into the intercellular spaces for feeding the fungal hyphae. The exposed fungal hyphae increase the surface of the root to several times. They perform several functions for the plant :

i. Absorption of water,

ii. Solubilisation of organic matter of the soil humus, release of inorganic nutrients, absorption and their transfer to root,

iii. Direct absorption of minerals from the soil over a large area and handing over the same to the root. Plants with ectomycorrhiza are known to absorb 2-3 times more of nitrogen, phosphorus, potassium and calcium,

iv. The fungus secretes antimicrobial sub-stances which protect the young roots from attack of pathogens. Ectomycorrhiza occurs in the trees like Eucalyptus, Oak (Quercus), Peach, Pine, etc. The fungus partner is generally specific. It belongs to basidiomycetes.

(b) Endomycorrhiza (- Endotrophic Mycorrhiza)

Fewer fungal hyphae lie on the surface. The remaining live in the cortex of the root, mostly in the intercellular spaces with some hyphal tips passing inside the cortical cells, e.g., grasses, crop plants, orchids and some woody plants. In seedling stage of orchids, the fungal hyphae also provide nourishment by forming nutrients rich cells called pelotons. Intracellular growth occurs in order to obtain nourishment because unlike ectomycorrhiza, the cortical cells do not secrete sugars in the intercellular spaces.

The hyphal tips passing into cortical cells either produce swollen vesicles or finely branched masses called arbuscules. Therefore, endomycorrhiza is also called VAM or vesicular-arbuscular mycorrhiza. The major benefits of VAM to the plant are the supply of inorganic nutrients as well as enhanced water absorption. Phosphate which is mostly present in the unavailable form in the soil, becomes abundantly available to the plant. A single fungus may form mycorrhizal association with a number of plants, e.g., Glomus.

Importance of Biofertilizers

i. They increase the yield of plants by 15-35%.

ii. Bio-fertilizers are effective even under semi-arid conditions,

iii. Farmers can prepare the inoculum themselves,

iv. They improve soil texture,

v. Bio-fertilizers do not allow pathogens to flourish,

vi. They produce vitamins and growth promoting bio-chemical's,

vii. They are non-polluting.

□□□

Chapter 8

Soil Management

As mentioned in the chapter 1, there are different soils in India and elsewhere, some of which require special management options for reaping higher farming benefits. Among them, vertisols, saline and alkali soils, calcareous soils, acid soils and acid sulphate soils are given with their properties and management.

Vertisols

Properties

Vertisols are black soils, poorly-drained fine-textured clay soils, are generally found on sedimentary plains, both on level land and in depressions. They occur in climates ranging from sub-humid temperate and Mediterranean to semi-arid and sub-humid tropical, with marked dry and wet seasons. The distinctly seasonal rainfall ranges from 150 mm to 2000 mm but is generally between 500 and 1000 mm per annum. The largest area of vertisols occur in Africa (105 million hectares), Asia and the Far East (57.8 million hectares), and Australia (48 million hectares). Vertisols owe their specific properties to the dom inance of swelling clay minerals, mainly montmorillonite. They show a great uniformity in physical characteristics, since these are largely dictated by the high clay content (40-80 per cent) and by the specific clay mineralogy. In the dry season the soils develop wide and deep cracks. These cracks close when the clays swell after the first rains. The swelling causes tensions leading to internal mass movements in the soil (churning or pedoturbation). This causes a characteristic structure to develop, with wedgeshaped structural aggregates in the surface soil and large, slickensidefaced planar soil blocks lower in the solum. Under the poor drainage conditions that are common in vertisol regions, leaching of soluble weathering products is severely restricted, the soil pH is above 7, and there is much available calcium and magnesium. These conditions favour the formation of smectite-type clay minerals, notably montmorillonite.

In semi-arid areas, free carbonates and gypsum accumulations are common. Saline and sodic vertisols may develop under irrigation, but they are rare under natural conditions. Most vertisols have a rather high but unbalanced fertility status. Vertisols can produce crops year after year at a sustained level, without being fertilized or manured. This is because the pedoturbation continuously brings subsoil to the surface. Any nutrient deficiency problems can easily be corrected: nitrogen and phosphate will be low. Some fixation as tricalcium phosphate may occur, but it is far less than in the acid tropical soils (oxisols, ultisols) of wetter regions. Response to potassium is variable. Secondary elements and micronutrients are often deficient.

Vertisols differ in surface characteristics and these strongly influence their reaction to soil tillage operations. There are two broad groups: - the ***self-mulching vertisols.*** These have a fine (granular or crumb) surface soil, 2-30 cm thick, during the dry season. This fine tilth is produced by dessication and soil shrinkage. When such soils are ploughed, the clods, after being subjected to repeated wetting and drying, disintegrate. When this mulch is well developed, seedbed preparation is hardly necessary, crust in the dry season. Such soils require mechanical tillage if they are to be cultivated. - the ***crusty vertisols.*** These have a thin, hard The self-mulching versus crusting characteristic is related to the tensile stress of the soil. One of the factors that influence this stress is soil texture. Very high amounts of sodium favour the formation of a hard surface crust.

Problems

Vertisols have a very hard consistence when dry and they are very plastic and very sticky when wet. The optimum soil moisture range for tillage (moist soil with a friable consistence) is narrow and vertisols are rarely in this state for long. With the use of heavy machinery, however, tillage operations can also be performed in the dry season. Mechanical tillage in the wet season causes serious soil compaction. Dry vertisols with a surface mulch or fine tilth have a high infiltration rate. When the soil is wet, however, the clays swell, closing the cracks and surface macropores. The soil thus becomes almost impermeable. Vertisols have a very low hydraulic conductivity: there is practically no water movement once the soil has reached its field capacity. As moisture penetration is limited, the volume of soil in which water is stored is small. Flooding can be a major problem in areas with higher rainfall because stagnant water can hamper tillage operations. The surface water can be drained by open drains, or crops can be grown on ridges and the intervening furrows used to direct the excess water to a main collector drain. Mole drainage is virtually impossible. As vertisols have a low structural stability they are very susceptible to water erosion. Slopes above 5 per cent should therefore not be used for arable cropping, and on gentler slopes contourcultivation with a groundcover crop is advisable, or the land can be used for pasturage. When terracing, sufficient surface drainage must be provided to avoid slumping.

Successful forest plantations have been reported, but in general tree crops do not come up well on vertisols: they lean and their roots can be broken when large cracks develop in the soil. In addition, the low subsoil porosity in the wet season discourages root development. The adverse physical properties of vertisols have been a major obstacle to agricultural land-use in low-technology societies: these soils are 'heavy' in the true sense of the word and are very difficult to work with hand-powered implements. Saline and sodic vertisols may develop under irrigation when the irrigation water is of poor quality. Once such a situation exists, soil improvement becomes very difficult.

Saline and Sodic Soils

Of the various problem soils in the world, saline and sodic soils occupy nearly 400 m.ha area. They are mostly confined to arid and semi-arid climates, but also occur in humid climates, especially in coastal areas. Australia has the largest area of saline and sodic soils, followed by Africa and Latin America.

Saline soils are defined by the electrical conductivity of saturation extracts of the soil horizons or by their soluble salt contents. The salt content may fluctuate, depending on evaporation and rainfall, or on irrigation history. Sodic (formerly called 'alkali') soils have at least 15 per cent exchangeable sodium that may decrease their physical stability. Many have a poor structure and a very low hydraulic conductivity. Most are alkaline, with a pH above 8.5 owing to the presence of sodium carbonate. In acid sodic soils and in sodic vertisols the 15 per cent limit is unsatisfactory because these soils do not show evidence of the characteristic low stability and poor structure except at appreciably higher proportions of exchangeable sodium.

Problems

The salts in saline soils or exchangeable sodium in sodic soils hinder crop growth. For efficient crop production they must therefore be leached from the root zone. This procedure is itself problematical because in most regions with these soils, irrigation water is scarce. Some soils that are more difficult to reclaim may require several times as much irrigation water *as* others. If the land is not carefully levelled before basin irrigation is carried out,. any variations in the microrelief (of the order of 5-10 cm) will cause considerable differences in the amounts of irrigation water entering the soil. Saline and sodic patches may thus persist or develop rapidly, even in land originally non-saline.

Most of the water applied to saline soils percolates through and removes the salts. However, if evaporation is high or the hydraulic conductivity of the soil is low, much more water must be applied to ensure percolation. Savings can be made if the irrigation water can be applied during seasons with a low evaporation rate. Alternatively, the sustained infiltration rate may be increased by measures such as cultivation, deep ploughing, gypsum application, or a combination of these. A further problem arises if the irrigation water is saline. In principle, the sodium carbonate present in many sodic soils can be leached: However, the associated high proportion of exchangeable sodium can only be leached after it has been replaced by other cations. These may be supplied in different ways, for example from gypsum dissolving in the percolating water, or from calcium carbonate or clay minerals dissolved in acid. These amendments (i.e. calcium salts or acids) may be added to the irrigation water or applied to the soil surface. More water is required to leach a sodic soil than a saline soil. This is because the usual chemical amendments are less soluble than the usual salts and because sodic soils are generally less permeable than saline ones. If the percolation rate is much lower than the evapotranspiration rate, the amendment may have to be incorporated into the surface horizon and more applied superficially. Sodic soils with a silt loam or silt texture and a homogeneous system of very fine interstitial pores have a very low saturated conductivity and leaching water barely moves through them under gravity. Water moves predominantly upwards under the influence of evaporation, and therefore any salts and sodicity accumulate near the surface.

Soils with a low capillary conductivity (e.g. clayey soils) can usually be drained satisfactorily by installing a system of ditches or pipe drains, 1.5 to 2.5 m deep. However, soils with a higher capillary conductivity (e.g. silty soils) must be drained deeper to prevent salinization or sodification, unless the groundwater quality is good. Deep drainage in areas receiving a strong inflow of groundwater from adjacent regions poses economic problems.

Characteristics of saline soils : Saline soils contain neutral soluble salts of chlorides and sulfates of sodium, Potassium, calcium and magnesium. The electrical conductivity of saturated extract of the soil is more than 4 dS m^{-1}.The ESP is less than 15 and pH is less than 8.5. Because of the presence of excess salts and low amount of Na+, these soils are in a flocculated state and their permeability is higher than alkali soils. Their physical condition is good and water can pass through them. These soils have a white crust of salts on their surface. Salinisation refers to the accumulation of neutral soluble salts in soils.

Characteristics of saline - alkali soils: These soils have both soluble salts and exchangeable sodium. The soil reaction becomes strongly alkaline because of hydrolysis of exchangeable sodium. The physical condition of the soil is deteriorated. The EC of saturated extract is more than 4 dS m^{-1}, the ESP is more than 15 and pH is around 8.3 to 8.6. They are transitional soils that they may be converted into saline or into sodic soils.

Characteristics of alkali or sodic soils: Most of the Na+ is in exchangeable form. EC of saturated extract is less than 4 dS m-1. ESP is more than 15 and pH as high 10. Such soils are not in a good physical condition and the surface of these soils is sometimes black due to dispersion of organic matter and humus. Such soils if ploughed when wet, turn into slick furrow slice referred to as slick spots.

Formation of saline and alkali soils

1. **Weathering of rocks and minerals** : The soluble salts released during weathering leach down with rain water, but again move upward due to high temperature and accumulate as white crust forming a saline soil.

2. **Hydrolysis of sodium saturated exchange complex in the formation of an alkali soil** : Sodium carbonate undergoes hydrolysis to form sodium hydroxide, which dissociates to increase concentration of hydroxyl ions in soil solution.

   Clay Micelle]Na + $H_2CO_3 \rightarrow$ Clay Micelle]Na + H + Na_2CO_3

   Na_2CO_3 + 2 H.OH $\rightarrow$ 2 NaOH + H_2CO_3

   2 NaOH $\rightarrow$ 2 Na^+ + OH

   Formation of OH- results in an increase in pH.

3. **Salty underground water** : If the ground water is salty, the salts move upward with water and get deposited when water evaporates.

4. **Arid and semi arid climate** : The low rainfall is not sufficient to leach down the soluble weathered products and hence the salts accumulate in the soil.

5. **Quality of irrigation water :** If water contains excessive amounts of Ca^{2+}, Mg^{2+}, Na^+, Cl^-, HCO_3^-, $CO_3^=$, BO_3^{3-} their continuous application develops salt affected soils. Besides this, a rise in the groundwater level, poor drainage which prevents leaching of salts and continuous use of bad quality irrigation water lead to the secondary salinisation and alkalisation.

6. **Use of basic fertilizer** : Basic fertilizers like $NaNO_3$ and basic slag may develop soil alkalinity.

Nutrient availability in saline and alkali soils

Presence of excessive quantities of salts in saline soils result in an increase in osmotic pressure of soil solution (osmotic effect). This results in the movement of water from plant cell into the concentrated soil solution. Alkali soils adversely affect plant growth in many ways.

i. Such soils are in poor physical condition. The presence of high Na+ keeps the soil particles suspended.

ii. Soils contain excessive amounts of boron, bicarbonates and sodium.

iii. Active sodium ions interfere with plant nutrition and metabolism. It adversely affects Ca^{2+} and Mg^{2+} nutrition of plants.

iv. Due to high pH, availability of phosphorus and micronutrients becomes low; the soils encounter Fe, Cu, Zn and Mn deficiencies.

v. Plant suffers oxygen deficiency as soil structure is adversely affected due to the dominance of sodium.

Reclamation of Salt Affected Soils

Before proceeding for reclamation of salt affected soils, the following points should be considered and available information should be collected on

i) Nature and extent of salinity.
ii) Quality of irrigation water with respect to sodium, boron and bicarbonate and salt content.
iii) Drainage conditions.
iv) Nature and distribution of salts in the root zone.
v) Calcium carbonate content of the soil.

I. Mechanical amelioration of salt affected soils

The commonly followed physical or mechanical method of amelioration of salt affected soils include deep ploughing, sub soiling, sanding, profile inversion and scraping.

The first two methods break the impermeable layer, hard pan or cemented sub soil layer existing at various depths in soil profile to improve the internal drainage of the soil and to facilitate the transportation of salts dissolved in water to deeper layers.

Incorporation of sand in salt affected soils is done to bring about permanent changes in texture, increase permeability and to improve water relations in the root zone.

Profile inversion can be adopted only under conditions where surface soil is good but the soil below is sodic or saline.

Scraping is adopted to remove the few centimeters of salt encrustation.

II. Chemical amelioration

A. Reclamation of a saline soil

Leaching : The main objective in reclamation of these soils is to leach the salts below the root zone (hence, drainage system should be installed if necessary). Leaching requirement (LR) is the fraction of water that must be leached through the root zone to control soil salinity at a specified level. This is achieved by flooding and draining. To make it effective, bunds are raised around plots prepared and water is applied depending on their water requirement to leach salts.

Leaching requirement (LR) = ECaw / EC dw

ECaw is the EC of applied water.

ECdw is the EC of drainage water.

In saline soils with high water table artificial drainage should be practiced (Drainage is the removal of excess water from the soil).

B. Reclamation of alkali (Sodic) soil

Reclamation of sodic soils involve two stages. First, the replacement of sodium by another cation and second, the leaching of the desorbed sodium salts out of the root zone. This may be done with gypsum ($CaSO_4 . 2 H_2O$) and leaching with good quality irrigation water.

2 Na – x + Ca2+ (solution) → Ca – x + 2 Na+ (solution)

Where x is the soil exchange complex

Gypsum requirement : Gypsum requirement is the calculated amount of gypsum necessary to add to reclaim the soil. It is the amount of gypsum required to be added to a sodic soil to lower its ESP (Exchangeable Sodium Percentage) to a desired level.

$$\text{Calcium requirement} = \frac{\{\text{ESP (initial)} - \text{ESP (desired)}\} \times \text{CEC}}{100} = \text{me of Ca/ 100 g soil}$$

1 me of Ca = 20 mg of Ca or 172 mg of Gypsum

ESP (Initial) is the ESP of soil before application of gypsum

ESP (desired) is the ESP of soil after bringing it to required level.

ESP in soil below 15 is safe for any crop. Suppose if the ESP is 25, it is required to bring this level to fifteen by removing 10 % of Na occupied on exchange sites of clay (CEC say 60 me / 100 g soil). So, in this situation 6 me of Na has to be removed to bring it below 15 %. To remove this sodium, 6 me of Ca is required i.e. 516 mg of Gypsum is required per 100 g soil. In one hectare land with BD of 1.33, the soil quantity will be around 2.25 miillion kgs. Hence, nearly 12 tons of Gypsum is required to remove 6 me of Na.

$$2\ NaHCO_3 + CaSO_4 \rightarrow CaCO_3 + Na_2\ SO_4 + CO_2$$

$$Na_2\ CO_3 + CaSO_4 \rightarrow CaCO_3 + Na_2SO_4$$

$$\text{Clay Micelle] Na} + CaSO_4 \rightarrow \text{Clay Micelle] Ca} + Na_2SO_4$$

This should be followed by application of good quality water to leach salts. Rate of gypsum application depends on soil pH and the amount of exchangeable sodium present on the soil exchangeable complex. Gypsum is suitable for alkali soils upto a pH of 9.0. Elemental sulphur or pyrites are effective on alkali soils which are calcareous (contain calcium carbonate) in nature.

$$2\ S + 3\ O_2 \rightarrow 2\ SO_3 \text{ (microbial oxidation)}$$

$$2\ SO_3 + 3\ H_2O \rightarrow 2\ H_2\ SO_4$$

$$H_2\ SO_4 + CaCO_3 + H_2O \rightarrow CaSO_4. + 3\ H_2O + CO_2$$

Here H_2SO_4 produced also reduces soil pH.

III. Biological amelioration

Organic materials and the activity of plant roots improve biological activity in the soil. During the decomposition of organic materials CO_2 is released which forms carbonic acid thus dissolving calcium compounds. This can be accomplished by greenmanuring, incorporation of crop residues, application of FYM, pressmud and other organic materials.

Classification of crops based on salt tolerance plant tolerance to salt concentration is due to

1. Accumulation of high level of sodium and chloride in shoots
2. Exclusion of salts by root cells,
3. Excretion of adsorbed salts from the plant by means of salt glands as in halophytes.

Salt Tolerance of Crops

1. Highly tolerant eg : *Sesbania*, rice, sugarcane, oats sugarbeet, Turnip, Date palm, ber.

2. Moderately tolerant eg : Castor, cotton, sorghum, maize, wheat, spinach, guava, pomegranate.
3. Moderately sensitive eg : Radish, cabbage, tomato, sweet potato.
4. Highly sensitive eg : Carrot, onion, lemon, orange, grape, apple, pulses, sesamum, pea.

Green manure crop is to be raised and incorporated at flowering stage into the soil immediately after reclamation. Rice crop is preferred to be grown after the green manure crop owing to its high tolerance to soil sodicity.

Management of Acid Soils

Acid soils can be managed by either growing crops suitable for particular soil pH or by ameliorating the soils through the application of amendments which will counteract soil acidity. Acid soils are made more suitable for agricultural use by liming which raises the soil pH. Liming increases the exchangeable base status, influences nutrient uptake, reduces toxic concentration of aluminium and manganese by neutralizing effect, improves the soil structure and promotes root distribution.

$$CaCO_3 + H_2O + CO_2 \rightarrow Ca\ (HCO_3)_2$$

$$Ca\ (HCO_3) \rightarrow Ca^{2+} + 2\ HCO^{3-}$$

$$H\text{-}Soil + Ca^{2+} \rightarrow Ca - soil + 2\ H^+$$

$$H+ + HCO_3- \rightarrow H_2CO_3 \rightarrow H_2O + CO_2$$

The higher the soil moisture, the rapid is the rate of reaction.

Lime requirement of acid soil : is the amount of a liming material that must be added to raise the soil pH to some prescribed value usually in the range of 6.0 to 7.0. The buffer method was proposed by Shoemaker et al. (1961). Liming material is the material which contains Ca and Mg that Can neutralize soil acidity. Lime stone, marketable lime, lime shells, paper mill sludge and basic slag are some of the liming materials available in the market. The efficiency of liming materials can be judged on the basis of the following factors.

i) Neutralizing value (N.V) or $CaCO_3$ equivalent (CCE) is defined as the acid neutralizing capacity of an agricultural liming material expressed as a weight percentage of calcium carbonate.

CCE of liming material = Molecular weight of $CaCO_3$ x100

Mol. Wt. of a liming material

The $CaCO_3$ equivalent of burnt lime is calculated as Molecular weight of $CaCO_3$ x100

Mol. Wt. of Cao = 100/56 = 1.786

ii) Purity of liming material: The more purer the material the higher will be the effectiveness.

iii) Degree of fineness of liming material: The rate of reaction of liming materials with an acid soil depends upon the fineness of the material.

Classification of Crops Based on Acidity Tolerance

1. Slight tolerance (upto pH6.5): eg.,Alfalfa and sugar beet.
2. Moderate (pH 6.0): eg., Peanut, cabbage, carrot, onions, radish, spinach and cauliflower.
3. Medium (pH 5.5) : eg., Cowpea, soybean, cotton, wheat, oats, peas, tomato, sorghum and millets.
4. Strong (pH 5.0): eg., Tobacco, apple, grapes, plum, watermelon
5. Very strong (pH 4.5) : eg., Citrus.

Problems associated with over liming: When excessively large amounts of lime is applied to an acid soil, the growth of the plant is influenced by one or many of these causes

Deficiency of iron, copper and zinc will occur.

Phosphorus and potassium availability is reduced.

Due to high OH- concentration by over liming, root development will be inhibited in association with tip swelling brought about by hydrations.

Due to dehydrating properties of boron, it acts as a protective agent for excess OH- concentration.

Due to over liming, boron deficiency occurs.

$$2\ Al\ X_3 + 3\ CaCO_3 + 3\ H_2O \rightarrow 3\ CaX_2 + 2\ Al(OH)_3 \downarrow + 3\ CO_2$$

Where 'X' is the exchange site. The freshly formed Al(OH)3 is then available for adsorption of boron thus causing boron deficiency.

Due to the application of lime in excess, the incidence of diseases like scab in root crops will be increased.

Due to over liming, the soils may become calcareous.

All these effects can be reduced by the application of large amounts of organic manures like FYM, incorporation of green manure crops.

Calcareous Soils

Calcareous soils are soils in which a high amount of calcium carbonate dominates the problems related to agricultural land use. They are characterized by the presence of calcium carbonate in the parent material and by a calcic horizon, a layer of secondary accumulation of carbonates (usually Ca or Mg) in excess of 15% calcium carbonate equivalent and at least 5% more carbonate than an underlying layer. These soils are formed under arid or semi-arid climatic conditions when the carbonate concentration in the soil solution remains high. In some soils the calcium carbonate deposits are concentrated into layers that may be very hard and impermeable to water (also called "Caliche"). These caliche layers are formed by insufficient rainfall (at nearly constant annual rates) leaching the salts to a particular depth in the soil at which the carbonates precipitate. They are also formed by salts moving upward from a water table (caused by irrigation) and precipitating near the top of the capillary fringe. The total extent of calcisols is estimated at 800 million hectares worldwide mainly concentrated in arid or Mediterranean climates. The land use of calcareous soils is highly variable: it ranges from non-used wastelands (deserts) to intensively cultivated irrigated areas.

Main production constraints

Calcareous soils develop in regions of low rainfall and must be irrigated to be productive. Therefore one of the main production constraints is the availability of water for irrigation. The quality of the irrigation water is of crucial importance for sustainable agricultural production on calcareous soils. Crusting of the surface may affect not only infiltration and soil aeration but also the emergence of seedlings. Cemented conditions of the subsoil layers may hamper root development and water movement characteristics. Calcareous soils tend to be low in organic matter and available nitrogen. The high pH level results in unavailability of phosphate (formation of unavailable calcium phosphates as apatite) and sometimes reduced micronutrient availability e.g. zinc and iron (lime induced chlorosis). There may be also problems of potassium and magnesium nutrition as a result of the nutritional imbalance between these elements and calcium.

Management

- Addition of large amounts of organic matter which releases organic acids during decomposition for solubilising $CaCO_3$ and to release the fixed P.
- Regular application of $ZnSO_4$ @ 25 – 50 kg ha^{-1}
- Application of phosphatic fertilisers by placement method to overcome the problem of P fixation.
- Foliar application of micronutrients in their respective sulphate forms.

Acid soils : Soils with pH < 5.5 have been defined as acid soils by USDA. However, soils with pH < 6.5 can also be categorized as acidic soils. Out of 157 M ha of cultivable land in India, 49 M ha are acidic, of which 26 M ha land is having a pH of less than 5.6 and the rest 23 M ha of land is having a soil pH in the range of 5.6 to 6.5.

Formation : The leaching of bases is the prerequisite for the formation of acid soils which are dominantly found to occur in regions with high rainfall. The major process involved in the formation of acid soils is podzolisation in areas of temperate climate; laterization of varying degrees, marshy conditions with significant amounts of partly decomposed organic matter. Acid soils occur in almost all major soil groups except the black soils (Vertisol). Acid soils occur in Assam, Manipur and Tripura, peaty and marshy soils of West Bengal and Kerala. Very low pH (4 or less) is an indication of the presence of sulfuric acid as in the cat clays in Kerala, the soils are referred to as acid sulfate soils.

Characteristics and Nutrient availability

Low pH, high exchangeable H^+ and Al^{3+}, low CEC and high base unsaturation are the characteristics of acid soils. Adverse effect is due to toxic concentration of Al, Mn and Fe and deficiency of Ca and Mg. Acid soils are low in available P and have high P fixing capacity. Available micronutrient status is adequate except molybdenum. The population of bacteria and actinomycetes is lower and those of fungi higher. Rock phosphate is the suitable P fertilizer than soluble phosphate in acid soils.

Acid Sulphate Soils

Acid sulphate soils have predominantly well developed soil profiles with characteristics of Sulfic Tropaquepts. They developed in pyritic (1-2.5% sulphur) non calcareous gray clays, deposited in brackish tidal environment 3000-7000 years ago. More recent fluviatile and marine deposits of adjacent upstream and downstream parts of the Plain gave rise to soils without acid sulphate conditions, of higher productive potential, and characteristics of predominantly Typic Tropaquepts.. The relation between soil units and soil productivity has been established on the basis of mainly morphometric criteria and statistical data correlated during surveys of land and soil. These can be ameliorated by combined liming and P-fertilization.

❑❑❑

Chapter 9

Irrigation Water

Quality of water is an important consideration in any appraisal of salinity or alkali conditions in an irrigated area. Salinity is one of the most important criterion for assessing quality of irrigation water especially under arid and semi arid region. Ground water contributes to about 40 % of all irrigation. The quality of irrigation water depends on the amount and kind of salt dissolved in it.

Harmful effects of poor quality irrigation water : If water containing excessive soluble salts is applied to the field, the soil turns saline to alkali. Harmful effects due to poor quality water through increased osmotic pressure of soil solution, thus inhibiting uptake of water and nutrients from soil to plant roots. Excessive sodium adsorbed by clay causes deflocculation, destroys soil structure resulting in poor circulation of air and water in the soil.

A. Classification of irrigation water or criteria for judging the quality of irrigation water given by USSL

1. Classification Based on EC (Electrical Conductivity)

C1 – Low salinity water, EC < 250 mho/cm. This water can be used for most crops

C2 – Medium salinity water, EC ranges between 250-750 mho/cm. Can be used with moderate leaching.

C3 – High salinity water, EC between 750 to 2250 mho/cm. This water cannot be used on soil with restricted drainage.

C4 – Very high salinity water, EC > 2250 mho / cm. Not suitable for irrigation.

2. **Classification based on SAR** (Sodium Adsorption Ratio) SAR = *Na / (Ca+Mg)* ½

   The concentrations of Na^+, Ca^{2+} and Mg^{2+} are expressed as meq L^{-1}

   S1 – SAR < 10. Low sodium water can be used on all soils with little danger of development of harmful level of exchangeable sodium.

   S2 – SAR is 10-18. Medium sodium water, can produce hazards with fine textured soils which have high CEC especially with low leaching conditions.

   S3 – SAR is 18-26. High sodium water; may produce harmful level of exchangeable sodium in most soils. It requires special management practices such as good drainage, high leaching and organic matter additions. Exchangeable Na should be replaced by the use of chemical amendments.

   S4 – SAR > 26. Very high sodium water, not suitable for irrigation. This water can be used with the addition of gypsum or other amendments.

3. **Classification based on RSC (Residual Sodium Carbonate)**

   RSC me L^{-1} = $(CO_3^{2-} + HCO_3^-) - (Ca^{2+} + Mg^{2+})$

   RSC value (me L^{-1}) Water quality

   < 1.25 Water can be used safely

   1.25 – 2.5 Water can be used with certain management

   > 2.5 Unsuitable for irrigation.

4. **Boron Concentration**

   It is evident that boron is essential for normal growth of the plant, but the amount required is very small. The permissible levels of boron in irrigation water is 1 ppm.

5. **Chloride Concentration**

   It is evident that chlorine is essential for normal growth of the plant. The permissible levels of chlorine in irrigation water is 10 ppm.

Use of Saline Water in Agriculture

The change in quality of irrigation water is technically not feasible and not economically viable. The following measures should be adopted for the use of saline water.

1. Selection of salt tolerant crops and varieties.
2. Deep ploughing to break the hard pan of salts if any.
3. Furrow planting is best for saline conditions because the seed can be safely planted below the zone of high salt accumulation.
4. Poor germination, high mortality of young seedlings and poor tillering are common features when the crop is grown with poor quality water. So, higher seed rate and close spacing is advisable. An additional seed rate of 25% should be adopted.
5. Dhaincha as a green manuring crop improves physical properties of the soil.
6. Addition of organic manures to some extent mitigates the adverse effect of poor quality water.
7. Fertilizers should be applied 1.25 - 1.5 times the normal rate of their application.
8. Application of Zn @ 20 kg $ZnSO_4$ ha^{-1} counteracts the negative effect of higher salinity and sodicity.
9. Split application of nitrogen to prevent N losses through volatilization and denitrification. Correction of nutrient deficiencies by foliar application of nutrients.
10. Dilution and cyclic use of good and saline waters. When good quality water is limited, it can be used as follows :
    - Presowing and first irrigation should be with good quality water. Later saline water can be used.
    - Poor quality water can be mixed with good water.
    - Drip or pitcher irrigation is found suitable.
11. Use of mulches to reduce the requirement of water for evaporation.

    Use of mulches and intercultural operations reduce water requirement of crops, thus with saline water salinity develops at a relatively lesser intensity.

Salt tolerance of different crops

High tolerance or Highly tolerant: Rice, cotton, sugarbeet, tobacco, date palms, ber.

Moderate tolerance : Spinach, sugarcane, wheat, guava, pomegranate, tomato, sweet potato.

Highly sensitive : Beans, pea, grape, orange, apple, pear, carrot.

Treatments with amendments

Water of high RSC can be used for crops having low water requirement. The rain water should not be allowed to leave the field as runoff but should be allowed to percolate through the soil profile.

Use of gypsum : In water containing excess of sodium (when RSC > 4 me L^{-1}), gypsum can be successfully used. With the addition of gypsum, the adverse effects of sodium can be mitigated.

Gypsum can be applied by

a) Keeping gypsum lumps or powdered gypsum in bags in small tanks constructed at the mouth of the tube well or water passage.

b) Application of gypsum, @ 75 kg ha^{-1} for each irrigation of 5 cm depth for 1 me L^{-1} of RSC above 2.5 me L^{-1}, should be applied to the soil.

c) A saturated solution of gypsum may be added at a constant rate to the flowing water.

$$2\,NaHCO_3 + CaSO_4 \;---\; Ca\,(HCO_3)_2 + Na_2\,SO_4$$

$$Ca\,(H\,CO_3)_2 \;---\; CaCO_3 + H_2O + CO_2$$

$Na_2\,SO_4$ being soluble can be leached.

Water containing toxic amounts of boron can be reduced by mixing with gypsum. Sodium borate is converted to calcium borate which being soluble can be leached out of root zone.

❑❑❑

Chapter 10

Remote Sensing and GIS

Remote Sensing

The soils are studied now-a-days from space. The fast developing technology that is being used in soil science is the Remote Sensing. It is the acquisition of information about an object or phenomenon without coming into physical contact. It refers to the use of aerial sensor technologies to detect and classify objects on Earth (both on the surface, and in atmosphere and oceans) by means of propagated signals (e.g. electromagnetic radiation).

Remote sensing of earth's environment comprises of measuring and recording electromagnetic energy reflected by the earth surface into the atmosphere and relating to surface conditions and atmospheric conditions. Sensors mounted on aircraft or satellite measure the energy reflected. It may be split into active remote sensing (when a signal is first emitted from aircraft or satellites) or passive (e.g. sunlight) when information is merely recorded. Active sensors work in micro-wave region of electromagnetic spectrum (radiation wavelength >1mm to 100 mm). Solar sensors operates in visible (0.4 to 0.7 micro meter) and infrared (1 to 100 micro meter) regi*on* of electromagnetic spectrum. Basic assumption made in the remote sensing is that an object has characteristic reflectance of incident radiation. Spatial reflectance curves describes the spectral responses of an object which depends on sun's orientation and elevation angle, sensors look angle and nature of the object.

In microwave radiation sensing method, back scattering coefficient (BSC) will be worked out in decibels (dB). Higher dB indicates rough surface. Calm Sea surface appear dark in Synthetic Aperture Radar. Trees and vegetation have rough surface, so appear bright. The tropical rainforest have BSC of -6 and -7 dB. The brightness of areas covered by bare soil may vary from dark to very bright depending on its roughness and moisture content. This depends on spatial resolution which is the smallest object that can be distinguished by the sensor from a point. The spectral resolution is the width of the spectral band and the number of spectral bands in which the image is taken. Narrow band width and more in number allows distinguishing between various features more easily.

Since 1970s multi-spectral platforms such as Landsat have been in use, which are obtained from Earth observation satellites, including the Landsat program satellite. Maps of land cover and land use from thematic mapping can be used for minerals, detect or monitor land usage, vegetation, and examine the health of indigenous plants and crops, including entire farming regions or forests. Landsat images are used by regulatory agencies to indicate water quality parameters, chlorophyll density and total phosphorus content. Weather satellites are used in meteorology and climatology.

Indian Remote Sensing Satellite

Indian Remote Sensing Satellite (IRS) mission takes up the projects of getting information on natural resources. Series of satellites, IRS 1A, 1B, 1C, 1D and P4 are launched by the Govt. of India. The cameras with LISS-II of 36.25 m resolution and LISS-I of 72.5 m resolution are used in the satellites. The panchromatic cameras on IRS1D provide data with spatial resolution of 5.2 to 5.8 m (at looking angle) and ground swath of 63 km to 70 km (at looking angle, i.e. at nadir). The camera operates in 0.5 to 0.75 microns band width. The LISS-II camera provides multi spectral data collected in four bands of visible, near infra red and short wave infra red regions. The spectral resolution is 21.2 to 23.5 m and with a swath of 127 to 141 km.

Normalized Difference Vegetation Index (NDVI)

The normalized difference vegetation index (*NDVI*) is a simple graphical indicator that can be used to analyze remote sensing measurements, typically but not necessarily from a space platform, and assess whether the target being observed contains live green vegetation or not.

The NDVI is calculated from these individual measurements as follows:

NDVI = (NIR-VIS) / (NIR+VIS)

where VIS and NIR stand for the spectral reflectance measurements acquired in the visible (red) and near-infrared regions, respectively.

These spectral reflectances are themselves ratios of the reflected over the incoming radiation in each spectral band individually, hence they take on values between 0.0 and 1.0. By design, the NDVI itself thus varies between -1.0 and +1.0.

This FCC (False Color Composite) images will be utilized in GIS. The ENVI (Environment for Visualizing Images) provides processing tools for land cover analysis.

Geographic Information System (GIS)

It is a computer system for capturing, storing, sharing, displaying, managing and analyzing the data and associated attributes which are spatially referenced to the earth.

Arc GIS and ERDAS

The software's are required to study the data input and satellite imagery (FCC) viz., Arc GIS, ERDAS, Surf and many more. In these softwares there are two methods for representation of geographical entities, 1.Raster method and 2. Vector method. In Raster, the terrain is divided into units of triangle, rectangle or hexagons. It is easy to use rectangle as grid cells. Here a number representing the entity for the grid will be given. In Vector, geographic space can be represented by the spatial location of attributes which are stored in another file for later use. The ground truth information is collected by using GPS (Global Positioning System) with latitude and longitude coordinates along with altitude. Our earth is not round, it is bulged at the equator and flat at poles. So, from west to east 20 grids of 6 degree each have been made (grids numbered from 1-60 left to right) and in north south direction 20 grids of 8 degrees have been made (grids numbered from C to X bottom to top with an exclusion of I and O). The grid lines of latitude are parrelel to equator and longitudinal lines are perpendicular. The coordinates of these grid points will be obtained by the GPS. The GPS data will used for providing the exact Raster and Vector to an attribute. By giving some colour and number to the attribute, the areas of same attribute will be taking the same colour. The different colour index can be shown on the area/map which can be printed as output. The soil fertility maps are obtained by data input of various soil parameters.

□□□

References

Alexander, M., 1977. Introduction to Soil Microbiology. Second edition, Publisher : John Willey & Sons, New York, 467 pages

Anji Reddy, M., 2006. Remote Sensing and Geographical Information System. 3rd Edition, BS Publications, Hyderabad, 453 p

Baver, L.D., W.H. Gardner, and W.R. Gardner, 1998. Soil Physics.4 th ed. Publisher: John Wiley & Sons, New York, 308 p

Biswas, T.D. and Mukharjee, S.K., 2001. Text Book of Soil Science. Tata McGraw-Hill Education., New Delhi, 433 p

Brevik, E.C., Hartemink, A.E., 2010. Early Soil Knowledge and the Birth and Development of Soil Science. *Catena* 83: 23–33.

Dilipkumar Das, 1996, Introduction to Soil Science. Publisher :Kalyani Publisher, New Delhi, 642 pages.

Firman E. Bear, 1964. Chemistry of the Soil. Second Edition, Publisher: Oxford & IBH Publishing Co. Pvt. Ltd., New Delhi, 515 p

Jackson, M.L., 2014. Soil Chemical Analysis, Revised edition, Publisher : United Book Prints, New Delhi.

Kim H. Tan, 2009. Principles of soil Chemistry. Fourth Edition - Publisher :CRC Press Book. New Delhi, 215 p

Nyle C. Brady, 1195. Nature and Properties of Soils. 10th Edition, Printice Hall of India, New Delhi, 621 pages

Piper, C.S., 2005. Soil and Plant analysis. Publisher : Srishti Book Distributors, New Delhi.

Sehgal, J., Text book of Pedology : Concepts and Applications. Publisher : Kalyani Publisher, New Delhi

Soil Survey Staff , 1993. Soil Survey : Early Concepts of Soil. *Soil Survey Manual USDA Handbook* 18, Soil Conservation Service. U.S. Department of Agriculture.

Sheela Devi L., Ramakrishna Parama, V.R., and Mruthunjaya, S., 2003. Nutrient Management of Plantation and Spice Crops. Krishi Vikas Publications, Bangalore, pp. 248.

Tisdale, S.L. and Nelson, W.L. and Beaton, J.D., 1990. Soil Fertility and Fertilizers. Fourth edition, Publisher : MacMillan Publishing Company, New York, 753 pages.

□□□

About the Authors

Dr. G.R.Rajakumar, Assistant Professor of Soil Science, Department of Natural Resource Management, College of Forestry, Sirsi was born on 28.05.1968 in Chitradurga District, Karnataka, India. He obtained his B.Sc.(Agri.) degree from College of Agriculture, Raichur, Karnataka during the year 1991, M.Sc. (Agri) in Soil Science & Agricultural Chemistry from UAS, Dharwad, Karnataka during the year 1994 and Ph.D in Soil Science & Agricultural Chemistry from UAS, Dharwad during the year 2000 with good academic records. He has undergone National training on Soil Health Management at JNKVV, Jabalpur during the year 2014. He has an experience for over 18 years in teaching (College/KVK) / research (ARS/College/KVK) and extension (KVK/College). He has published 12 scientific full length papers in National and International Journals as author / co-author, and 6 papers presented in Symposium as author/co-author, written 3 books and 24 popular articles and released 3 technologies as co-principal investigator in package of practices book of UAS Dharwad.

Dr. Sangappa V. Patil, Professor of Agronomy, Regional Horticultural Research and Extension Centre, Bengaluru, was born on 01.06.1969 in Gulbarga District, Karnataka, India. He obtained his B.Sc.(Agri.) degree from College of Agriculture, Raichur during the year 1991, M.Sc. (Agri) in Agronomy from UAS, Dharwad during the year 1994 and Ph.D in Agronomy from UAS, Dharwad during the year 2011 with good academic records. He has undergone International training on Potato for a period of 10 weeks in IAC, Wageningen, Netherland during the year 2003. He has an experience for over 22 years in teaching, research and extension. He has to his credit about 48 scientific full length papers published in National and International Journals as author/co-author, about 26 papers presented in National and International Symposium as author/co-author and 6 books. Eleven technologies have been released as co-principal investigator in package of practice book of UAS Bangalore and UHS Bagalkot. He has been awarded as Best Farm Superintendent for the year 2004 by Hon'ble Governor Shri T. N. Chaturvedi, Govt. of Karnataka during 39th convocation of UAS Bangalore held on 21st March 2005, Fellow of Hind Agri-Horti Society Muzafarnagar during the year 2011 and Best Agriculture Scientist 2014 by Lingaraja Vividhoddeshagala Samsthe, Bharati Nagar, Dharwad, Karnataka. He is awarded by the prestigious organization 'Venus International Foundation (R)', Chennai as the 'Out Standing Faculty Award 2015 in the field of Agronomy' for his achievements in the career.